Access to Building Services Engineering

Levels 1 and 2

Diane Canwell
Peter Marini
Neil McManus
Chris Payne
Jon Sutherland

Nelson Thornes

Published in 2012 by:
Nelson Thornes Ltd
Delta Place
27 Bath Road
CHELTENHAM
GL53 7TH
United Kingdom

12 13 14 15 16/ 10 9 8 7 6 5 4 3 2 1

A catalogue record for this book is available from the British Library

ISBN 978 1 4085 1534 1

Cover photographs: Catherine Veulet/iStockphoto and Branks Miokovic/iStockphoto

Page make-up and Illustrations by GreenGate Publishing Services, Tonbridge, Kent

Printed and bound in Spain by GraphyCems

Acknowledgements

The authors and publisher would like to thank the following for permission to reproduce their material:

Buck and Hickman: Fig. 8.11 212167 Record 235C Chain Wrench © 2012 Buck and Hickman; **Crown Copyright**: Fig. 1.6 Part G of the Building Regulations cover Crown Copyright one HM Government Building Regulations © Crown Copyright 2010. Reproduced under PSI licence no. C2009002012; **ECA**: Fig. 1.8, ELECSA logo © 2011 ECA Certification Ltd.; **Energy Saving Trust**: Fig. 3.1, Energy Saving Trust logos © Energy Saving Trust 2011; **Fotolia**: Fig. 1.1, 1.4, 1.5, 1.9, 1.10, 1.11, 1.12, 2.1, 2.3, 2.4, p34, 2.7, 2.8, 2.9, 2.10, 2.11, 2.16, 2.19, 3.12, 3.13, 3.14, 4.2, 4.5, 4.8, 5.1, 5.2, 5.4, 6.1, 6.5, 7.6, 7.18, 7.26, 7.27, 8.6; **iStockphoto**: Fig. 1.2, 2.17, 3.11, 4.3, 5.3, 7.3, 7.4, 7.5, 7.7, 7.8, 7.9, 7.12, 7.13, 7.20, 7.21, 7.25, 8.5, 9.5; **NAPIT** Fig. 1.7, NAPIT logo © NAPIT 2007; **Neil McManus**: Fig 7.19, 7.32; **Rigid**: Fig. 8.8, Rigid 34955 Pipe Reamer Model 2-S © 2012 RIDGID. All rights reserved; **Science Photo Library**: Fig. 2.6 (Peter Gardiner); **Toolbank.com**: Fig. 7.15, Steel Conduit Bender © Toolbank.com 2000-2012; **instant art**: p26, p30, Fig. 2.2, 2.14; and **Heather Gunn Photography**.

Contents

Introduction

The Access to Building Services Engineering qualification is an introduction to the following industries:

- Refrigeration and air conditioning
- Electrical installation
- Heating and ventilation
- Plumbing

This book has been designed to cover the content of the Level 1/Level 2 qualification and includes activities to help you prepare for the assessments. It will introduce you to the basic systems, the skills required, and the tools, materials and components used in each industry. Step-by-step guides, some with photographs, guide you through the basic skills you will need.

How to use this book

Key terms: explanations of important terms that you will need to know and understand.

Trade tips: tips that will help you work more efficiently.

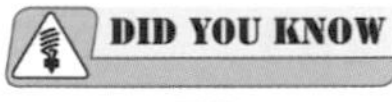

Did you know: interesting facts about the industry.

Toolbox talk: tips to help you to work safely.

Have a go: practical tasks to develop your skills and techniques in preparation for your assessment.

Activity: exercises to help you test and apply your knowledge.

Remember: important points to remember.

Quick quiz: multiple-choice questions to test your knowledge and provide practice for your assessment.

Indicates information only needed for the Level 2 Diploma.

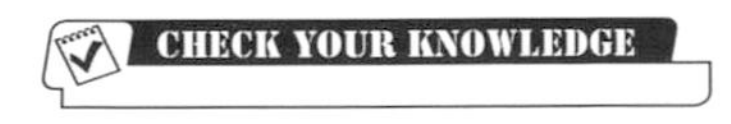

Check your knowledge: revision questions to test your knowledge and understanding of the chapter.

1 Introduction to Building Services Engineering

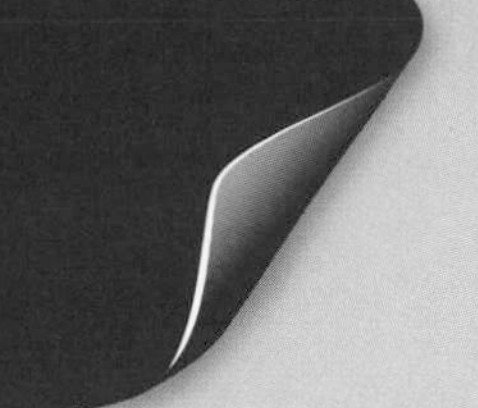

This chapter covers the learning outcomes for:

Introduction to Building Services Engineering

City & Guilds unit number 104; EAL unit code QACC1/04; ABC unit code A01.

Building services engineers install, service and maintain a huge range of different systems. They are also increasingly at the cutting edge of new technology. Many are involved in energy efficiency, and all play an important role in ensuring that, in an increasingly complex society, the safety, comfort and welfare of customers and clients is achieved.

In this chapter you will learn more about building services engineering – the different career opportunities available, as well as its impact on people's lives and the built environment. You will discover the range of systems within this sector, and the key laws and codes of practice that underpin it.

IN THIS CHAPTER YOU WILL LEARN ABOUT:

- the main industries
- key job functions in building services
- building services engineering systems
- legislation, standards and codes of practice.

The main industries

There are four main industries within the building services engineering sector:

- Plumbing – dealing with cold and hot water, heating and sanitation
- Electrical – dealing with power and lighting circuits
- Heating and ventilation – dealing with cold and hot water systems, heating, ductwork and specialist work, such as heat emitters and more complex heating systems
- Refrigeration and air conditioning – dealing with both refrigeration and air conditioning systems and units

Fig. 1.1 A plumber at work

Plumbers and electricians often begin their careers as apprentices.

Plumbers

Most plumbing businesses are relatively small. They may concentrate on particular types of plumbing, such as domestic work, installation or working with developers on new builds.

Some larger plumbing businesses work in most areas of plumbing, including heating and ventilation systems. They may also carry out electrical work. Other plumbing businesses deal only with companies or work on large commercial contracts.

For most plumbers, the nature of their work demands that they are highly mobile and available for emergency call-outs. For example, a domestic plumber typically will work at several sites over the course of a day. By contrast, plumbers working for larger businesses may work on bigger projects, in a single location, for a number of weeks.

Electricians

Electricians are involved in a wide range of activities. Without the wiring that they install, there would be no lights, computers or televisions. They install equipment to control how the electricity flows and are often involved in fixing domestic electrical equipment.

Many electricians focus on new builds, using **blueprints** to plan a circuit and install the electricity supply. Other electricians focus on specific installation projects, such as installing telephones, fire alarms, computers or **fibre optic cables**.

Electricians work wherever there is a need for electrical wiring and related equipment. Many operate as small businesses, while others work for larger electrical contractors on a full-time or job-by-job basis.

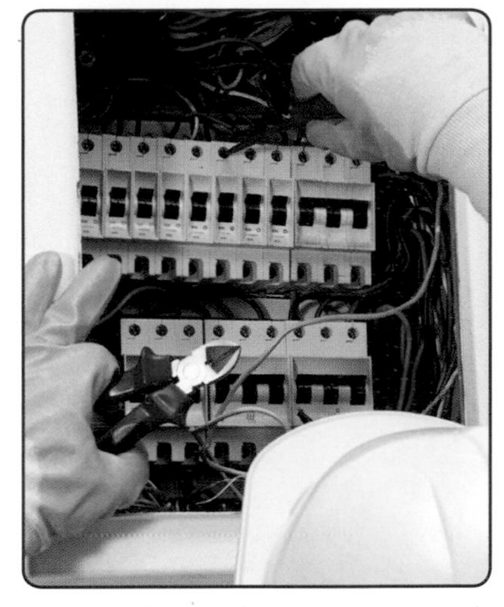

Fig. 1.2 An electrician at work

KEY TERMS

Blueprint: A design plan or technical drawing.

Fibre optic cable: A bundle of threads made of pure flexible glass, used instead of metal wiring.

Commission: A process in which the installer checks the systems components and overall performance before it is handed over to the building owner and signed off.

Contract specification: To formally agree upon the requirements of a job to be done.

Heating and ventilation engineers

Heating and ventilation engineers work on central heating systems, and ventilation systems that keep dwellings or buildings cool. They install, **commission** and maintain complex systems for both domestic projects and major building sites.

Heating engineers are often involved in the design, **contract specification**, and inspection and testing of new installations, as well as upgrading existing systems. They are also involved in gas appliance work.

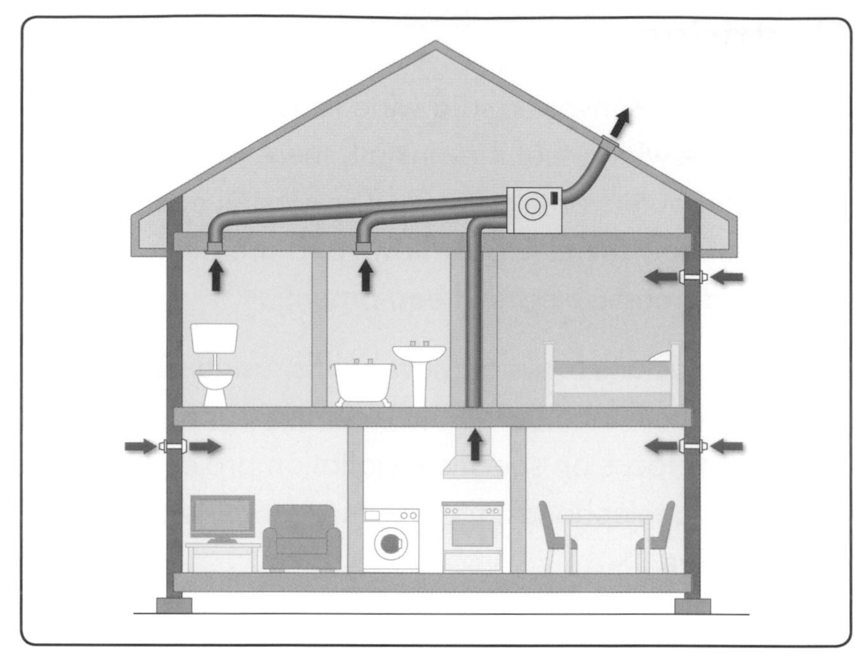

Fig. 1.3 Heating and ventilation is about keeping a building at the right temperature in summer or winter

DID YOU KNOW

Ventilation replaces air in order to control temperature or to remove odours, heat or dust. Air conditioning further controls temperature by using heat or refrigeration.

Refrigeration and air conditioning engineers

Refrigeration and air conditioning systems use mechanical, electrical and electronic components. They control temperature, humidity and air quality.

Refrigeration systems make it possible to store and transport perishable items.

Refrigeration and air conditioning engineers specialise in installation, maintenance and repair. They work from blueprints and specifications to install the systems, the ducts and the vents.

Fig. 1.4 Repairing an industrial air conditioning unit

Refrigeration engineers often work on:

- commercial refrigeration systems
- connecting motors

- compressors
- condensing units and evaporators
- programme control systems.

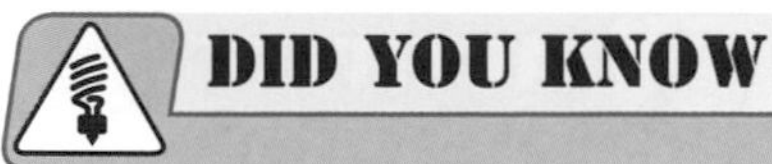

The Institute of Refrigeration is a professional body for refrigeration and air conditioning engineers.

Building services sector – everyday life and buildings

Buildings and their construction have a considerable impact on the environment. They account for half of all energy consumed and produce half of all greenhouse gases. This means that building services engineers are now looking to install natural ventilation systems and use renewable sources of energy in both domestic and commercial sites. For example:

- solar energy is used to power air conditioning systems
- buildings are better insulated and made more airtight, to help conserve energy.

Building services make the inside of a building comfortable and safe. This includes:

- the supply of energy – electricity, gas and renewable energy
- heating and air conditioning
- water supply, plumbing and drainage
- artificial lighting
- lifts and escalators
- telephone and IT networks
- alarm and security systems
- fire detection and protection systems.

TRADE TIP

All building services engineers are often required to work evenings and weekends, e.g. to complete urgent jobs or emergency work.

ACTIVITY

For each situation below, which type of building services engineer would be involved?

1. A hospital operating theatre needs humidity and bacteria controls.
2. An intensive care unit needs to be able to strictly control the heating.
3. A forensic laboratory needs a foolproof air filtration system.

Fig. 1.5 It starts with the plan

There is a demand that buildings do far more than simply provide shelter – they have to offer a comfortable, safe and modern environment.

The construction and built environment sector

The construction and built environment sector is extremely diverse. In effect, people working within the sector fall into three main categories:

- Designers and consultants – those that plan the buildings
- Suppliers – those that provide raw materials and components
- Contractors – those that carry out the construction.

Building services engineers are, therefore, a main part of building, construction and property services, which also include:

- architects
- designers and planners
- builders
- civil engineers
- demolition specialists
- restorers
- suppliers and manufacturers of products.

Across the UK, there are over 1 million people working in all branches of construction and building services engineering. This includes around 194,000 **private contractors**, who range from individuals working alone, all the way up to businesses that employ over a thousand people.

Key job functions in building services

Many of the key trades in building services engineering provide almost guaranteed work throughout the year. They can also provide excellent career pathways.

These trades may give you the opportunity to become **self-employed** and to run your own business. This means dealing with all aspects of the business yourself, including paying taxes and advertising the company. As with many service-based occupations, success often depends on your reputation and receiving recommendations from satisfied customers.

Few building services careers can promise predictable working hours. This is particularly the case if you provide services to domestic customers. If you decide to work for a larger business, you will tend to work more regular hours.

Private contractor: A person or company that is temporarily employed by another business or an individual.

Self-employed: Someone who is in business for themselves and provides a service to a number of different customers or clients.

Gas fitter

Job functions	Qualifications and career progression
A gas fitter installs, maintains and repairs gas lines in dwellings, as well as in commercial or industrial sites. They will regularly respond to calls regarding leaks and malfunctions. Natural gas can be a hazardous energy source, so efficiency and precise work are essential. As with many professions, there is rapid technological change. New hi-tech equipment can make the job challenging.	Some courses to become a gas fitter can be completed in a relatively short time. Some gas fitters are also fully qualified plumbers. It is important to be a registered gas fitter and to keep the licence up to date. A gas fitter will also need to be on the Gas Safe Register.

Domestic plumber

Job functions	Qualifications and career progression
Some plumbers are 'wet only' and handle bathrooms, kitchens and radiators. Others are 'gas only' and focus on boilers and heating systems. Plumbers maintain water piping and discharge pipework. This includes drinking water, heating, venting and sewerage. A good plumber would have to be able to: ■ read drawings ■ install, repair and maintain a range of fixtures and systems ■ locate pipe connections, passage holes and fixtures ■ often work in confined spaces or at height ■ work out time and cost estimates for jobs ■ test pipes for leakages ■ meet rigorous safety standards and Building Regulations.	Many domestic plumbers begin as apprentices, during which they undergo NVQ training. Additional qualifications can help focus on specific plumbing areas, such as gas work. A Master Plumber Certificate is the next step. To install gas heating systems, a plumber needs to be on the Gas Safe Register and pass an Approved Certification Scheme. You must also demonstrate competence on a range of items including oil, solid fuel, unvented hot water (HW) and water regulations.

Industrial and commercial plumber

Job functions	Qualifications and career progression
Rather than working in a domestic environment, these plumbers focus on working for businesses or in other locations. Some are self-employed and others are employees of larger plumbing companies. Some work in factories, hospitals or laboratories, fitting, maintaining and repairing specialist equipment.	Working as an apprentice or a 'mate' and job shadowing a trained plumber is often a good start. Larger plumbing businesses have more established apprenticeship systems or schemes. Many plumbers work for smaller businesses until they are experienced, qualified or confident enough to work alone.

Service and maintenance engineer

Job functions	Qualifications and career progression
These engineers provide planned maintenance services to meet the individual client's requirements. These can relate to plumbing, ventilation, heating or air conditioning. The work is a mix of planned maintenance services and emergency call-outs due to breakdowns.	An apprenticeship leading to qualification in service and maintenance engineering is necessary. Career progression depends on your chosen specialism and the range of heating and ventilating systems and equipment that can be tackled.

Maintenance electrician

Job functions	Qualifications and career progression
Maintenance electricians tend to work in larger businesses such as factories or hospitals. They are responsible for keeping the lighting, electrical systems and generators in working order. They need to diagnose problems and then carry out repairs. A great deal of their time is spent on preventative maintenance using periodic inspections.	Maintenance electricians will often begin their careers as apprentices. As with all other types of electrician, maintenance electricians need to be compliant with Building Regulations Part P (Electrical safety). They usually achieve this as part of a VRQ Certificate in the Building Regulations for Electrical Installations in Dwellings.

Installation electrician

Job functions	Qualifications and career progression
An installation electrician primarily fits new electrical components, but will also be involved in maintenance and repair. Some installers focus on wiring ships or aircraft. 'Installation electrician' describes electricians that deal primarily with domestic dwellings, so they would fit, test and repair electrical circuits and wiring.	It is essential to have NVQ qualifications in electrotechnical technology as well as knowledge of wiring regulations. Domestic electricians will also need to be compliant with Building Regulations Part P (Electrical safety). They usually achieve this as part of a VRQ Certificate in the Building Regulations for Electrical Installations in Dwellings.

Heating installer

Job functions	Qualifications and career progression
A heating installer will be involved in a variety of hot and cold water installations. They will organise installations of water heaters and boilers, including all pipework. They will also carry out inspection, maintenance and repair, including the correcting of faults.	Obtaining a Level 2 Technical Certificate in Heating, Design and Installation or Heating and Ventilating is a good start. The next step is to become a heating engineer on the Gas Safe Register. Many heating engineers are self-employed, although working with larger businesses will mean that training is funded.

Ductwork installer

Job functions	Qualifications and career progression
This job requires the installation of ductwork to: ■ carry warm, cold or humid air into a building ■ remove contaminated or stale air from a building. Much of the work is done on larger construction sites.	It is necessary to gain NVQ Level 2 or 3 in Mechanical Engineering Services (Heating and Ventilating). This qualification is normally gained during an apprenticeship. By the end of the apprenticeship, you would be a skilled installer and eligible for registration.

Refrigeration engineer

Job functions	Career progression
A refrigeration engineer installs and services refrigeration systems, including those in dwellings, commercial buildings and delivery vehicles. They follow blueprints and instruction manuals to install new systems, connecting necessary electrical wires, water and coolant pipes and air ducts.	It is necessary to complete an apprenticeship of two to four years, together with an NVQ Level 3 in Refrigeration and Air Conditioning. Many engineers work for larger businesses covering specific areas or types of refrigeration system. Larger projects are often carried out in teams.

Air conditioning engineer

Job functions	Career progression
These engineers design, install, repair and maintain systems in homes, offices, retail outlets or other public buildings. Their work focuses on regulating air quality and temperature. After installation, inspection tests and regular servicing of systems take place.	An apprenticeship combines training and on-the-job experience together with an NVQ Level 2 or 3 in Refrigeration and Air Conditioning. An HNC in Construction and the Built Environment will enable you to become an air conditioning contracts manager or building services engineer.

ACTIVITY

Go to www.plumbingjobsonline.co.uk and type in your intended career occupation.

1. What employment opportunities are there in your local area?
2. What is the average hourly or annual pay rate?
3. What are the minimum qualifications required for these job roles?

Building services engineering systems

You need to be able to:

- identify a range of basic systems across the different occupations
- state their basic functions and main components.

The table below gives a summary of the basic systems and the pages where you will find more detailed information about these systems.

Occupation	Basic system	Pages
Refrigeration and air conditioning	Refrigeration systems, air conditioning systems	154–155 and 161–164
Electricial	The power circuit (otherwise known as the ring main or radial), the lighting circuit	174–175
Heating and ventilation	Cold water, hot water, heating, duct work, specialist (e.g. chilled water)	206–208
Plumbing	Cold water, hot water, heating and sanitation	237–239

Legislation, standards and codes of practice

Each part of the building services engineering sector is controlled, regulated or guided by legislation, standards and codes of practice. These set out:

- the minimum requirements of work carried out
- the expected behaviour and required levels of competence of people working in the sector.

Identifying legislation, standards and codes of practice

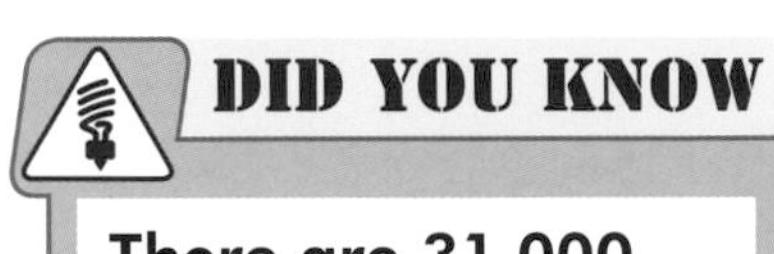

There are 31,000 British standards.

These three aspects always need to be considered when carrying out work in the building services engineering sector. The difference between them is the source of the guidance:

- *Legislation* is laws or regulations that have been approved by the British parliament, which aims to safeguard or govern the way work is done.
- *British and European standards (BS EN)* – goods and services are marked to show that they meet the requirements of specific standards within designated schemes. In the past, British standards used to be a series of numbers preceded by BS. Today, most standards are BS EN, which means they are both British and European standards. There are also ISO standards, which are international.
- *Codes of practice* set out the professional standards usually required by an organisation, for example the Association of Plumbing and Heating Contractors. They are designed to improve industry practices and to recognise highly skilled individuals.

DID YOU KNOW

The Health and Safety at Work etc. Act (1974) is an example of legislation.

Building Regulations

These are designed to ensure that all buildings, from design through to construction, are safe, healthy, accessible and sustainable. They aim to improve building controls and to ensure compliance with the law.

The new Building Regulations came into force in October 2010. These sections are particularly important:

- Part G – sanitation, hot water safety and water efficiency
- Part F – ventilation in buildings
- Part J – heat-producing appliances
- Part L – conservation of fuel and power
- Part P – electrical safety

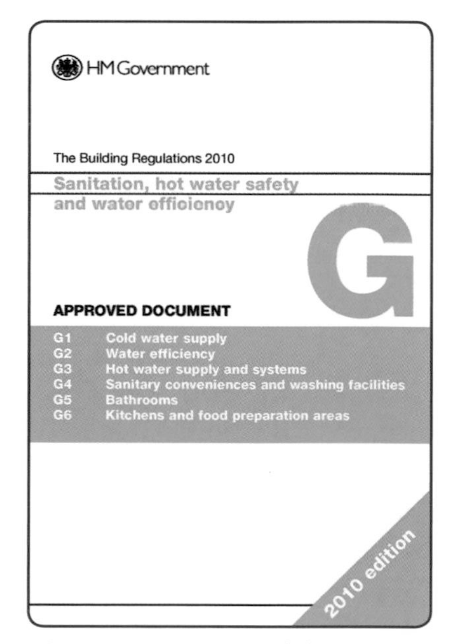

Fig. 1.6 Part G of the Building Regulations

Fig. 1.7 NAPIT (National Association of Professional Inspectors and Testers) is an example of a Competent Person Scheme

Consideration should also be given to Part A (Structural safety) and Part B (Fire safety), both of which can be affected by building services installation methods.

Competent Person Scheme

This allows individuals or businesses to self-certify that they comply with Building Regulations. It was introduced by the government to guarantee that work complies with the law.

The Competent Person Scheme covers many aspects of building work including:

- installation and replacement of roof coverings
- cavity wall insulation
- systems such as electrics and plumbing
- solid- and oil-fuelled burners and boilers and their associated hot water and heating systems.

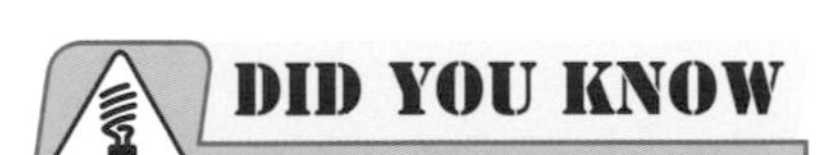

You can read a full list of all types of work covered by the Competent Person Scheme at: www.communities.gov.uk (document 09BD05896).

Health and Safety at Work etc. Act (HASAWA) (1974)

This piece of legislation aims to encourage businesses to manage health and safety risks, for example by providing appropriate training and facilities. It also covers first aid, accidents and ill health.

HASAWA applies to all building services engineers across the sector, whatever their work location. The Health and Safety Executive (HSE) is responsible for enforcing the act.

Fig. 1.8 ELECSA, owned by the ECA (Electrical Contractors' Association) is an example of a Competent Person Scheme specifically for electrical work

Reporting of Injuries, Diseases and Dangerous Occurrences Regulations (RIDDOR) (1995)

This legislation makes it a duty to report accidents, occupational diseases and 'near misses' to the HSE. For example, if you discovered a gas appliance or fitting that you considered to be dangerous, you would need to contact the HSE immediately via its Incident Contact Centre or website.

Fire Precautions (Workplace) Regulations (1997)
This requires employers to carry out a comprehensive fire risk assessment and to provide the necessary fire precautions, including means of detection and escape, and fire-fighting equipment.

Fig. 1.9 By law, all workplaces must have suitable fire-fighting equipment

Health and Safety (Safety Signs and Signals) Regulations (1996)
Whenever a risk has not been entirely eliminated or completely controlled, safety signs should be used. Any unfamiliar signs must be explained to employees and visitors to the site.

Provision and Use of Work Equipment Regulations (PUWER) (1998)
PUWER concerns health and safety risks related to equipment used at work. It states that any risks arising from the use of equipment must either be prevented or controlled, and all suitable safety measures must have been taken. In addition, tools need to be:

- suitable for their intended use
- safe
- well maintained
- used only by those who have been trained to do so.

Fig. 1.10 By law, employers must display appropriate safety signs

Electricity at Work Regulations (EWR) (1989)

These provide safety standards regarding the use of electricity. They require individuals and businesses to be responsible for the compliance of the regulations as far as 'is reasonably practicable'.

Fig. 1.11 All electrical work is governed by the Electricity at Work Regulations (EWR) (1989)

Fig. 1.12 PPE includes clothing and equipment to protect against risks in the workplace

Personal Protective Equipment at Work Regulations (1992)

This law states that employers must provide employees with personal protective equipment (PPE) at work whenever there is a risk to health and safety. PPE needs to be:

- suitable for the work to be done
- well maintained or replaced
- properly stored
- correctly used at all times, so employees must be trained how to use it appropriately.

British and European standards

There are several British and European standards that relate to all areas of construction work. They fall into seven different categories, as described in the table below.

Area of construction	Examples of British and European standards
Design	Mainly used by architects, design engineers, developers and main contractors – they cover areas such as building types, environmental engineering, concrete and plaster, plant and equipment
Structures	Mainly used by structural engineers – they cover foundations, piles and the use of steel and timber
Construction products	Mainly for manufacturers, inspectors and testers, covering concrete, responsible sourcing and other materials
Built environment engineering	This covers heating, ventilation, air conditioning and the installation of gas appliances, as well as electrical installation – it includes flueing and ventilation, and installation and maintenance
Infrastructure	These cover energy and water, such as sanitary installations, supply of water and roof drainage
Site operations	These cover mobile machinery, equipment, noise, environmental pollution and overall site safety
Facilities management	These cover the building maintenance and administration – they include emergency lighting and fire risk assessment

Codes of practice

Another part of the work of The British Standards Institution (BSI) is to create a range of codes of practice, or recommendations. These also fall into the seven categories, as shown in the table above.

BS EN have largely replaced the old BS (British Standards). The BS EN are European specifications.

Area of construction	Examples of codes of practice
Design	■ BS 8300 requires those designing buildings to meet the needs of disabled people ■ BS 9999 requires the design of the building to have acceptable levels of fire safety
Structures	■ BS 5268 covers the structural use of timber in building ■ BS 5950 covers the structural use of steelwork in building
Construction products	■ BS 8214 recommends specifications for fire doors ■ BS 5395 covers the design and building of staircases
Built environment engineering	■ BS 5440-1:2008 covers flueing and ventilation for gas appliances of rated input not exceeding 70 kW net ■ BS 5871-4:2007 covers specification for the installation and maintenance of gas fires, convector heaters, fire/back boilers and decorative fuel effect gas appliances
Infrastructure	■ BS 6465 gives recommendations on the selection, installation and maintenance of sanitary systems ■ BS 8515 covers rainwater harvesting
Site operations	■ BS 5228 covers noise and vibration control on sites ■ BS 7121 covers the use of cranes and lorry loaders
Facilities management	■ BS 5266 covers emergency lighting

DID YOU KNOW

Westminster City Council has a Code of Construction Practice at: www.westminster.gov.uk

In Scotland, a broader code can be found at: www.scottish.parliament.uk

Other codes of practice

Certain councils or boroughs also have their own codes of practice that they expect developers and contractors to follow. These will often include:

- demolition
- construction
- site clearance
- maintenance and repair.

Usually, a code of practice will apply not only to work being carried out for the authority, but to all work in the area. This is often referred to as a Code of Construction Practice.

CHECK YOUR KNOWLEDGE

LEVEL 1

1. Why are fibre optic cables used instead of older-style copper wiring on some projects?
2. What is the main difference between ventilation and air conditioning?
3. What type of components would refrigeration and air conditioning systems use?
4. Who would normally be involved in planning the look and construction of a building?
5. What are the main functions of a domestic plumber?
6. Which trade professionals might a heating installer work with on a regular basis?
7. Suggest three different types of location where an air conditioning engineer might work.
8. What are the alternative terms used to describe a power circuit?

2 Safe working practices

This chapter covers the learning outcomes for:

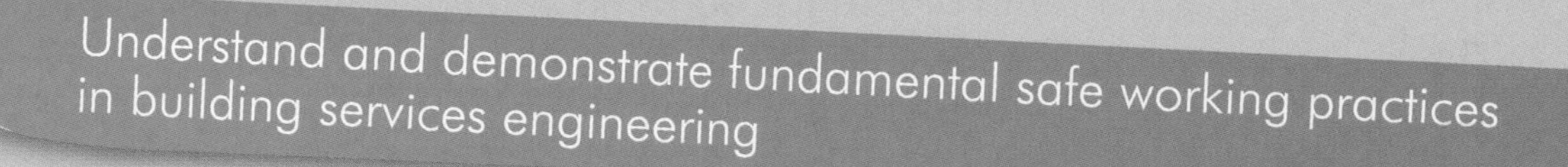

L2 Understand and carry out safe working practices in building services engineering

City & Guilds unit number 101 and L2 201; EAL unit code QACC1/01 and L2 QBSE2/01; ABC unit code AO2 and L2 AO6.

The construction industry can be dangerous, so keeping safe and healthy at work is very important. If you are not careful, you could injure yourself in an accident or perhaps use equipment or materials that could damage your health. Keeping safe and healthy will help ensure that you have a long and injury-free career.

Although the construction industry is much safer today than in the past, there are still thousands of people that are injured or killed every year. Many others suffer from long-term health problems, such as deafness, spinal damage, skin conditions or breathing problems.

IN THIS CHAPTER YOU WILL LEARN ABOUT:

- UK health and safety law
- hazardous situations
- asbestos
- personal protection
- manual handling
- accidents
- electrical safety
- access equipment
- heat-producing equipment
- excavations and confined spaces.

UK health and safety law

Laws have been created in the UK to try to ensure safety at work. Ignoring the rules can mean injury or damage to health. It can also mean losing your job or being taken to court.

The two main laws are the Health and Safety at Work etc. Act (**HASAWA**) and the Control of Substances Hazardous to Health (**COSHH**) Regulations.

KEY TERMS

HASAWA: The Health and Safety at Work etc. Act outlines your and your employer's health and safety responsibilities.

COSHH: The Control of Substances Hazardous to Health Regulations are concerned with controlling exposure to hazardous materials.

PPE: Personal protective equipment can include gloves, goggles and hard hats.

Health and Safety at Work etc. Act (HASAWA) (1974)

This law applies to all working environments and to all types of worker, sub-contractor, employer and visitors to the workplace. It places a duty on everyone to follow rules in order to ensure health, safety and welfare.

Employers

HASAWA states that employers with five or more staff need their own health and safety policy. Employers must assess any risks that may be involved in their workplace and then introduce controls to reduce these risks. These risk assessments need to be reviewed regularly.

Employers also need to supply personal protective equipment (**PPE**) to all employees and to ensure that it is worn when required.

DID YOU KNOW

In 2010–11, there were 52 fatal accidents in the construction industry in the UK. (Source: www.hse.gov.uk)

Employees and sub-contractors

HASAWA states that all those operating in the workplace must aim to work in a safe way. They must wear any

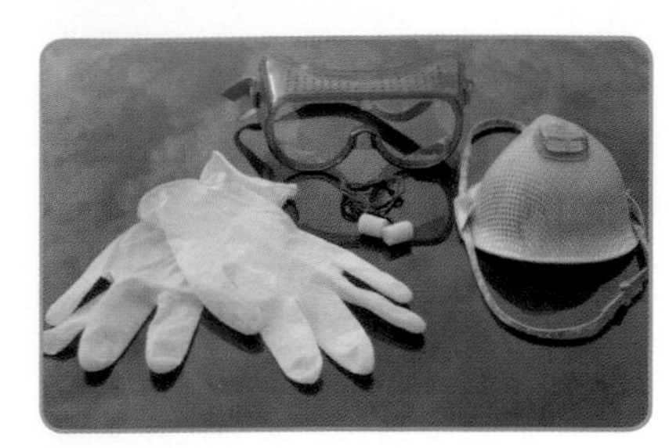

Fig. 2.1 Gloves, goggles and masks are types of personal protective equipment

KEY TERMS

HSE: The Health and Safety Executive, an independent organisation that ensures health and safety laws are followed.

Improvement notice: This gives the employer a time limit to make changes to improve health and safety.

Prohibition notice: This stops all work until the improvements to health and safety have been made.

Sub-contractor: An individual or group of workers that are directly employed by the contractor.

PPE provided and look after their equipment. Employees should not be charged for PPE or any actions that the employer needs to take to ensure safety.

The Health and Safety Executive (**HSE**) is an independent organisation that is responsible for health, safety and illness. It carries out spot checks on the workplace to make sure that the law is being followed.

The HSE has access to all areas and can bring the police if it thinks that there may be a major problem. It can also take away equipment. Should it find a problem then it can issue an **improvement notice**. This gives the employer a limited amount of time to put things right.

In serious cases, the HSE can issue a **prohibition notice**. This means all work has to stop until the problem is dealt with. An employer, the employees or **sub-contractors** could be taken to court.

QUICK QUIZ

What does PPE stand for?

a) Personal protective environment
b) Private and personal equipment
c) Protective preventative equipment
d) Personal protective equipment

Control of Substances Hazardous to Health (COSHH) Regulations (2002)

In construction, it is common to be exposed to substances that could cause ill health. For example, you may use oil-based paints or preservatives, or work in conditions where there is dust or bacteria.

Employers need to protect their employees from the risks associated with using hazardous substances. This means assessing the risks and deciding on the necessary precautions to take.

Any control measures have to be introduced and maintained, which includes monitoring an employee's exposure to harmful substances. The employer will need to carry out health checks and ensure that employees are made aware of the dangers and are supervised.

DID YOU KNOW

You can find out whether the CDM (Construction Design and Management) Regulations apply to where you work by going to: www.cdm-regulations-uk.co.uk

Construction (Design and Management) (CDM) Regulations (2007)

These regulations apply to safety on larger construction sites, and cover any type of work carried out on site. They aim to ensure that everyone involved in the project works in a coordinated way to ensure a safe environment. The regulations also address the need for regular inspections, procedures and maximum protection.

Reporting of Injuries, Diseases and Dangerous Occurrences Regulations (RIDDOR) (1995)

Under RIDDOR, employers are required to report any injuries, diseases or dangerous occurrences to the HSE. The regulations also state the need to maintain an **accident book**.

KEY TERMS

Accident book: This is required by law. Even minor accidents need to be recorded by the employer.

TOOLBOX TALK

Any accident at work that results in more than three days off work must be reported to the HSE.

Fire Precautions (Workplace) Regulations (1997)

These regulations state that employers need to carry out risk assessments to identify general fire precautions. These include:

- ways of detecting fire and giving an appropriate warning
- identifying escape routes
- positioning of fire-fighting equipment
- training employees in fire safety.

Health and Safety (Safety Signs and Signals) Regulations (1996)

These standardise the way of using safety signs. Employers need to use particular safety signs to help manage risks. They also need to maintain these signs and explain unfamiliar ones to their employees.

QUICK QUIZ

Which of the following laws require an employer to keep an accident book?

a) EWR Regulations
b) HASAWA
c) RIDDOR
d) CDM Regulations

Electricity at Work Regulations (EWR) (1989)

These regulations are designed to ensure that safety standards are maintained when dealing with electricity in the workplace. They apply to both employers and employees.

More information about electrical safety can be found later in this chapter (pages 47–51).

Responding to hazardous situations

On any construction site, there is always a risk of accidents occurring. Construction work is also often physically demanding, so there are potential health risks and chances of personal injury.

Typical construction accidents can include:

- falls from scaffolding, ladders and roofs
- electrocution
- injury from faulty machinery
- power tool accidents
- construction debris
- holes in flooring
- fires and explosions
- burns, including those from chemicals.

Safety notices

In a well organised working environment, there will be safety signs that warn you of potential dangers and tell you what to do to stay safe.

There are five basic safety signs, as well as signs that are a combination of two or more of these types. These are shown on the following table.

REMEMBER

Safety signs are there to inform you. Take notice of them for your own health and safety.

The different types of safety signs

Type of safety sign	What it tells you	What it looks like	Example
Prohibition signs	Tells you what you must *not* do	Usually round, in red and white	Do not use ladder
Hazard signs	Warns you about hazards	Triangular, in yellow and black	Caution Slippery floor
Mandatory signs	Tells you what you *must* do	Round, usually blue and white	Masks must be worn in this area
Safe condition or information signs	Gives important information (e.g. where to find fire exits, assembly points or the first aid kit), or about safe working practices	Green and white	First aid
Fire-fighting signs	Gives information about extinguishers, hydrants, hoses, fire alarm call points, etc.	Red with white lettering	Fire alarm call point
Combination signs	These have two or more of the elements of the other types of sign (e.g. hazard, prohibition and mandatory)		DANGER Isolate before removing cover

Personal conduct

How you behave – that is, your personal conduct – on a construction site is very important. You need to ensure that your own actions reduce the risks to yourself and to others.

You need to make sure that you DO NOT:

- perform any actions that could create a health or safety risk
- ignore risks in the workplace.

Instead, you should:

- take sensible action to put things right – this may mean reporting dangerous situations and seeking advice
- carry out working practices in line with legal requirements
- follow workplace policies
- pass on suggestions for reducing risks
- follow supplier and manufacturer instructions
- make sure you know how to use hazardous substances
- ensure that you know what to do in the event of an emergency.

Tools and equipment

All tools and equipment can be dangerous if they are misused, so it is important to always use the right tool for the job.

- Equipment such as drills and saws need to be sharp and kept in good condition. Never use saws with defective or worn teeth.

To maintain tools and keep them in good working condition, make sure they are cleaned regularly.

REMEMBER

Don't always rely on others to keep you safe! You can contribute to workplace safety by making sure that your working area is clear of hazards and is tidy.

- Check handles to make sure they are not broken or splintered.
- Check that power cables for tools are not frayed.
- Repair or replace damaged or broken ropes, buckets or barrows.

Trips and fire hazards

Leaving equipment and materials lying around can cause accidents, as can trailing cables and spilt water or oil. Some of these materials are also potential fire hazards.

Working practices

These three working practices can help to prevent accidents or dangerous situations occurring in the workplace:

- Method statements are used for high-risk activities. They summarise risk assessments and other findings to provide guidance on how the work should be carried out.
- Permit-to-work systems are used for very high-risk and potentially fatal activities. They are checklists that need to be completed before the work begins. They must be signed by a supervisor.
- Risk assessments look carefully at what could cause an individual harm and how to prevent this. The idea is that no one should be injured or become ill as a result of their work. Risk assessments identify how likely it is that an accident might happen and the consequences of it happening. A risk factor is worked out and control measures created to try to offset it.

Fire

Fires need oxygen, heat and fuel to burn. A spark can be the heat needed to start a fire, and anything flammable, such as petrol, paper or wood, provides the fuel.

In the event of a fire, raise the alarm quickly. Leave the building and head for the **assembly point**. Never stop to collect your belongings.

TOOLBOX TALK

Being tidy can help prevent fires from starting. Waste needs to be put into the proper place, such as a skip or bin. Dust is very flammable, so always use a dust bag.

Hazardous substances

COSHH Regulations identify a wide variety of substances that must be labelled in different ways.

Controlling the use of these substances is always difficult. Ideally, their use should be eliminated or substituted with something less harmful. Failing this, they should only be used in enclosed conditions. If none of this is possible then they should only be used in controlled situations.

If a hazardous situation occurs at work, then you should:

- ensure the area is made safe
- inform the supervisor, site manager, safety officer or other nominated person.

You will also need to report any potential hazards or near misses.

KEY TERMS

Assembly point: An agreed place outside the building to head for in the event of an emergency.

The six main categories of hazardous substances

Type of substance	Hazard symbol	How to prevent harm
Toxic		■ Close containers ■ Wear PPE ■ Wash hands
Harmful	h	■ Keep in sealed container ■ Use only as directed ■ May react with other substances, so avoid contact
Corrosive		■ Wear PPE ■ Flush with running water ■ Ensure good ventilation
Irritant	i	■ Do not breathe in the vapour or spray ■ Keep container tightly closed ■ Avoid contact with the skin and eyes
L2 Oxidising		■ Avoid eye or skin contact ■ Do not breathe in dust, vapour or spray
Extremely flammable		■ No flames or sparks ■ Use only in small quantities

Common substances

In addition to hazardous substances, there are various other substances you are likely to come across that could pose a risk. The following table tells you how to work with safely with these.

L2 *Safe ways to work with commonly encountered substances*

Substance	General precautions
Lead	■ Wear protective clothing and breathing equipment ■ Wash hands and face and scrub nails
Solvents and lubricants	■ Read suppliers' safety data sheets and labels ■ Work in a ventilated area and wear PPE
Fluxes	■ Wear PPE ■ Avoid skin contact ■ Attend health checks
Jointing compounds	■ Wear PPE, including splash-proof goggles ■ Ensure adequate ventilation
Sealants	■ Always check suppliers' instructions ■ Wear PPE
Gases	■ LPG and oxyacetylene are inflammable, so only use in well-ventilated areas and avoid sparks ■ Carbon dioxide can cause suffocation, so PPE and proper ventilation is essential
Cleaning agents	■ Check suppliers' instructions and chemical ingredients ■ Wear PPE as some can cause allergic reactions

ACTIVITY

1. Which of the commonly encountered substances in this table are you likely to come across in your chosen area of work?
2. How will you identify these substances, and how will you find out the general precautions that are necessary for working with them safely?

Asbestos

Asbestos is a mineral with heat-resistant properties that was used in many products and building materials. In older buildings, asbestos may be found in:

- insulating materials
- the building fabric

DID YOU KNOW

The three common types of asbestos are:

- **chrysotile (white)**
- **amosite (brown or grey)**
- **crocidolite (blue).**

- sheeting materials, floors, roofs and walls
- coating materials used to produce decorative finishes on walls and ceilings, such as Artex
- small gaskets and seals.

TOOLBOX TALK

Licensed contractors must always wear protective equipment. They make sure asbestos waste is double-bagged and wash their hands before breaks and leaving the site.

Breaking up materials that contain asbestos can create asbestos dust. If this is inhaled, it can cause serious diseases of the lungs, which may be fatal. The Control of Asbestos Regulations (2006), therefore, identifies asbestos insulating material as being a high-risk substance. Today, only licensed contractors are allowed to work with asbestos.

If you suspect that asbestos is present in a working area during a renovation project, you must stop work and seek guidance from the supervisor immediately.

Fig. 2.2 Asbestos warning sign

However, asbestos does not always need to be removed. Sometimes it may be better to leave it in place, in which case asbestos warning signs can be mounted on the material to alert people. In such cases, the presence, condition and form of presumed asbestos must be recorded.

Before you can work with asbestos, there must be a full **risk assessment**. This will establish the control measures to be taken. You would also have to attend an asbestos awareness training session.

DID YOU KNOW

The HSE has a full range of guidance notes on how to deal with asbestos and asbestos cement. For more information, go to: www.hse.gov.uk

All asbestos needs to be disposed of using proper packaging at approved waste sites.

Asbestos cement

Asbestos cement can be found in a wide variety of building materials. It is not high risk, but risk assessments and control measures still need to be put in place. For example:

- when working with flue, soil, rainwater pipes and gutters, you need to wear disposable overalls with a hood and single-use gloves
- when working with tanks and cisterns, you need to wear single-use gloves, disposable overalls with a hood and **respiratory protective equipment**.

KEY TERMS

Risk assessment: An investigation that highlights the risks involved in a job and how to deal with those risks. The findings are recorded.

Respiratory protective equipment: Masks and breathing apparatus designed to prevent inhalation of harmful substances.

Fig. 2.3 Removing materials containing asbestos from an old factory

Fig. 2.4 Asbestos cement roofing

ACTIVITY

Asbestos can damage your lungs. If asbestos is present, why should you not do the following?

- Use power tools.
- Reuse disposable clothing or masks.
- Eat or drink in the work area.

TOOLBOX TALK

Asbestos cement can be very fragile. There have been a number of deaths and serious injuries when people have fallen through asbestos roofs.

Personal protection measures

ACTIVITY

Identify each of the items of PPE that this man is wearing and state why he might be using each one.

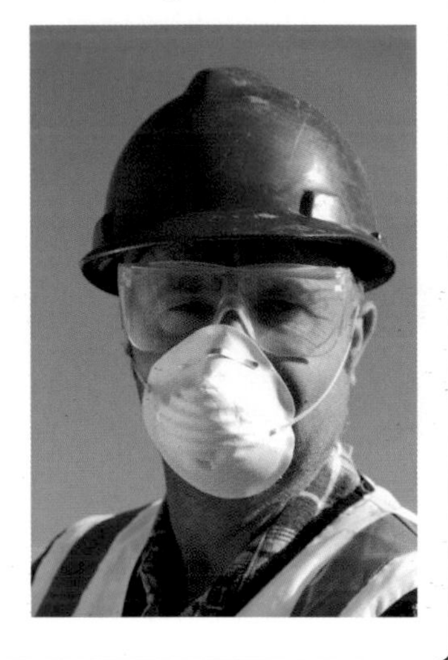

TRADE TIP

By law, employers must supply employees with high-visibility clothing free of charge.

PPE is used when an element of risk remains, even after all of the potential hazards have been minimised.

By law, employers need to:

- provide PPE free of charge
- ensure it is suitable for its intended use
- make sure it is cleaned, maintained and replaced if damaged
- provide a safe storage area for it
- make sure that it is used in the correct way.

Protective clothing

Protective clothing, such as overalls:

- provides some protection from spills, dust and irritants
- can help protect from minor cuts and abrasions
- reduces wear to work clothing underneath.

In certain circumstances, waterproof or chemical-resistant overalls may be required.

High-visibility clothing stands out against any background or in any weather conditions. It is important to wear high-visibility clothing on a construction site to ensure that people can see you easily. In addition, workers should always try to wear light-coloured clothing, as it is easier to see.

You need to keep your high-visibility and protective clothing clean and in good condition.

Employers need to make sure that employees understand the reasons for wearing high-visibility clothing and the consequences of not doing so.

Eye protection

Construction sites can be a hazardous environment for the eyes. This means it is essential to wear goggles or safety glasses to prevent small objects, such as dust, wood or metal, from getting into the eyes. As goggles tend to steam up, particularly if they are being worn with a mask, safety glasses can often be a good alternative.

Fig. 2.5 Safety glasses

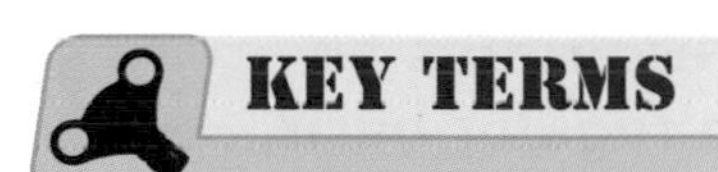

Contact dermatitis: Inflammation of the skin following contact with a particular substance. The skin becomes red, dry, itchy and sore.

Hand protection

Wearing gloves will help to prevent damage or injury to the hands or fingers. For example, general purpose gloves can prevent cuts, and rubber gloves can prevent skin irritation and inflammation, such as **contact dermatitis** caused by handling hazardous substances.

There are many different types of gloves available, including specialist gloves for working with chemicals.

DID YOU KNOW

A quarter of all work-related injuries are to hands and fingers.
Over 700,000 working days are lost every year due to contact dermatitis.

Head protection

Hard hats or safety helmets are compulsory on most building sites. They can protect you from falling objects or banging your head. They need to fit well and they should be regularly inspected and checked for cracks. Worn straps mean that the helmet should be replaced, as a blow to the head can be fatal.

REMEMBER

You may not pay much attention to loud noises, but they could be doing serious damage to your hearing.

TOOLBOX TALK

If you use respirators on a regular basis, make sure that you replace the filters or the respirator will not work properly.

QUICK QUIZ

You are using a noisy power drill to make a large hole in a concrete wall. Which of the following PPE should you use?

a) Ear protection
b) Dust mask
c) Goggles
d) All three above

Foot protection

Safety boots should have rubber soles to provide protection from electric shocks or sharp objects that you might stand on. They should also have steel toecaps to prevent injury if something falls on your foot.

Hearing protection

Ear defenders, such as protectors or plugs, aim to prevent hearing loss when you are working with loud tools or involved in a very noisy job.

L2 Respiratory protection

Breathing in fibre, dust or some gases could damage your lungs. Dust is a very common danger, so a dust mask, face mask or respirator may be necessary.

Make sure you have the right mask for the job. It needs to fit properly – otherwise it will give you no protection.

Manual handling

Procedures for manual handling

Lifting or handling heavy or bulky items is a major cause of injuries on construction sites. So whenever you are dealing with a heavy load, it is important to carry out a basic risk assessment.

The first thing you need to do is consider the job to be done and ask the following questions:

- Do I need to bend or twist?
- Does the object need to be lifted or put down from a distance?
- Does the object need to be carried a long way?
- Does the object need to be pushed or pulled for a long distance?
- Is the object likely to shift around while it is being moved?

If the answer to any of these questions is 'yes', you may need to adjust the way the task is done to make it safer.

The next thing that needs to be looked at is the object itself. Ask yourself the following:

- Is it just heavy or is it also bulky or an awkward shape?
- How easy is it to get a good handhold on the object?
- Is the object one item or are there parts that might move around and shift the weight?
- Is the object hot or does it have sharp edges?

Again, if you have answered 'yes' to any of these questions, then you need to take steps to address these issues.

It is also important to think about the working environment and where the lifting and carrying is taking place. Ask yourself:

- Are the floors stable?
- Are the surfaces slippery?
- Will a lack of space restrict my movement?
- Are there any steps or slopes?
- Is there enough light?

TOOLBOX TALK

The weight limit for lifting and/or moving heavy or awkward objects is 20 kg. Keep this in mind when manual handling and do not lift or move objects above this weight.

Safe lifting

Before lifting and moving an object:

- check that the pathway to where the load needs to be taken is clear
- look at the product data sheet and assess the weight. If you think the object is too heavy or difficult to move then ask someone to help you. Alternatively, you may need to use a mechanical lifting device.

When you are ready to lift, gently raise the load. Take care to ensure the correct posture – you should have a straight back, with your elbows tucked in, your knees bent and your feet slightly apart.

Once you have picked up the load, move slowly towards your destination. When you get there, make sure that you do not drop the load but carefully place it down.

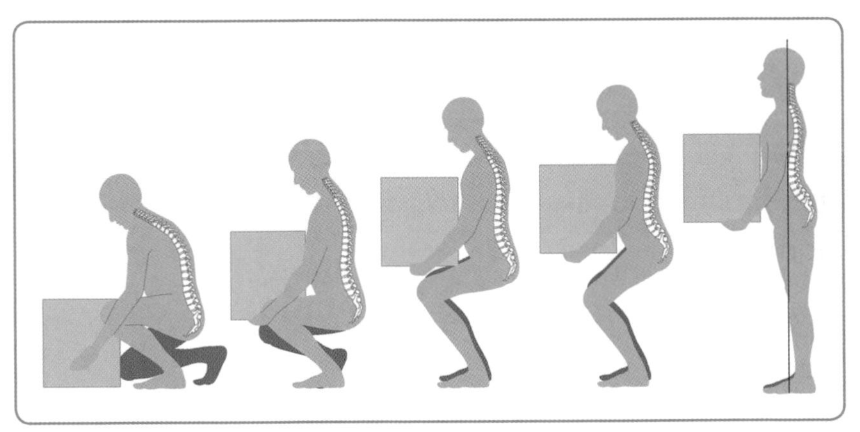

Fig. 2.6 Take care to follow the correct procedure for lifting

Mechanical lifting aids

Sack trolleys are useful for moving heavy and bulky items around. Gently slide the bottom of the sack trolley under the object and then raise the trolley to 45° before moving off. Make sure that the object is properly balanced and is not too big for the sack trolley.

On large construction sites, trailers and forklift trucks are often used, as are dump trucks. You should never use any of these unless you have been properly trained.

Fig. 2.7 Sack trolley

Fig. 2.8 Pallet truck

HAVE A GO...

Practice manual handling with a heavy and bulky item.

1. Plan the lift
2. Safely move the load
3. Repeat for a two-person lift

Can you explain why a two-person rather than a one-person lift is necessary?

Responding to accidents

The first thing to do if there is an accident is to raise the alarm. This could mean:

- calling for the first aider
- phoning for an ambulance
- dealing with the problem yourself.

Never move anyone who may have a back or neck injury, or who has fallen from a height.

DID YOU KNOW

The three main emergency services in the UK are:

- **the fire service (fire and rescue)**
- **the ambulance service (medical emergencies)**
- **the police (immediate police response).**

REMEMBER

The calmer you remain on the telephone, the quicker the emergency services will be with you.

Fig. 2.9 Assembly point sign

How you respond will depend on the severity of the injury.

You should follow this procedure if you need to contact the emergency services:

- Find a telephone away from the emergency.
- Dial 999.
- You may have to go through a switchboard. Carefully listen to what the operator is saying to you and try to stay calm.
- When asked, give the operator your name and location, and the name of the emergency service or services you require.
- You will then be transferred to the appropriate emergency service, who will ask you questions about the accident and the location. Answer the questions in a clear and calm way.
- Once the call is over, make sure someone is available to help direct the emergency services to the location of the accident.

Evacuation procedures

During an emergency, a general alarm will sound and you will need to make your way to a place of safety away from the site.

For larger and more complex construction sites, evacuation usually begins by only evacuating the area closest to the emergency. Areas are then evacuated one by one to avoid congestion of the escape routes.

Dealing with minor injuries

There may be times when you need to deal with minor injuries at work. Knowing how to deal with common accidents is important as even minor injuries could be dangerous. The following table shows the basic ways to deal with some minor injuries.

Approaches to some minor injuries

Type of injury	How to deal with it
Cuts	■ Wash and dry your hands ■ If you have any cuts on your own hands, put on some disposable gloves ■ Clean the cut under running water then pat dry with a sterile dressing ■ Cover the cut temporarily and clean the surrounding skin, and then re-cover with a sterile dressing or plaster
Minor burns	■ Cool the burn with cold or lukewarm water for 10–30 minutes ■ Remove clothing or jewellery from the injured area before it swells, but not if it is stuck to the burnt skin ■ Cover the burn with a layer of cling film (do not wrap around the limb), or use a clean, clear plastic bag for burns to the hands or feet (this will protect the burn from infection) ■ Never break the blister or apply ointment to the injured area
Objects entering the eye	■ Wash and dry your hands ■ Wash the eye with clean, cool water ■ Apply an eye pad and ensure the individual goes to hospital.
Exposure to fumes	■ Remove the casualty from further exposure if possible ■ Ventilate the area or remove the casualty to fresh air ■ Seek medical attention

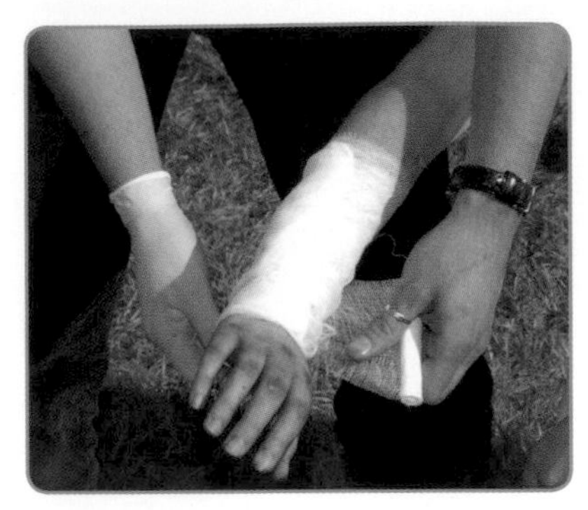

Fig. 2.10 Applying a sterile dressing

Fig. 2.11 A typical first aid box might contain sterile plasters, eye pads, triangular bandages and wound dressings

REMEMBER

A construction site with fewer than five people simply needs an appointed first aider. A construction site with up to 50 employees requires a trained first aider.

First aid

Sometimes it may be necessary for a first aider to give help. First aiders have been trained in first aid. They have to attend courses regularly to keep up to date.

If you are working in a small, occupied property then the minimum requirements are:

- a suitably stocked first aid box
- an appointed first aider
- for everyone to know what the first aid arrangements are.

If you are working on a construction site, then the HSE demands, in addition to the basic first aid requirements described above, that someone with a valid certificate of competence in first aid or emergency first aid is present. The type of training should reflect the type of injuries that might occur in that environment.

L2 Dealing with major injuries

Bone fractures

If you suspect that someone has broken a bone then take the casualty to hospital or call an ambulance. Do not move the casualty unless they are in danger of becoming further injured.

Responding to someone who is unconscious

If a casualty is unconscious they should be placed in the **recovery position**, unless they have a back or neck injury, in which case they should not be moved.

L2

Fig. 2.12 The recovery position

KEY TERMS

Recovery position: This involves rolling someone onto their side, with their head tilted back slightly to keep their airway open.

Concussion

Concussion is injury to the brain following an impact to the head. The patient may lose consciousness or feel dazed. If you suspect that someone is suffering from concussion, you must seek medical assistance.

All blows to the head could be dangerous and the brain will need time to heal.

REMEMBER

When dealing with a major injury, your first priority must be to seek medical help.

Electric shock

The extent of an electric shock will depend on the current and the length of time the person has been in contact with the electric circuit.

When responding to a person who has suffered an electric shock you must do the following:

1. **Make sure the person is no longer in contact with the electrical current source.** Do this by trying to switch off the power supply or pull out the plug, if possible. Alternatively, insulate yourself from the ground, for example with rubber matting, and

L2

use an object of low conductivity, such as a wooden broom or rolled-up newspaper, to push away the power source.

2. Check how conscious the person is. Talk to them and see if they are able to respond to questions.
3. If the casualty is not moving or breathing, call an ambulance immediately.
4. You may need to perform cardiopulmonary resuscitation (**CPR**), as described below.

KEY TERMS

CPR: Cardiopulmonary resuscitation – a combination of rescue breaths and chest compressions to keep blood and oxygen circulating through the body.

Performing CPR

CPR may be necessary in any major injury. Carrying out the ABC checks will tell you whether there is a need for CPR.

A Airways – open up the airway by lifting the chin and tilting back the head. Check for and remove any obstructions in the mouth or throat.

B Breathing – check for breathing by observing whether the chest rises and falls and by listening for breaths.

C Circulation – check for a pulse, for example, by placing two fingers over the inner edge of the wrist.

If the casualty is not breathing, CPR may be necessary. This involves mouth-to-mouth breaths followed by chest compressions.

Follow this procedure:

1. Open up the airway and remove any mouth obstructions.
2. Pinch the casualty's nose, take a full breath and blow into the casualty's mouth. You will see their chest rise.

REMEMBER

You will be putting yourself at risk of electric shock if you touch the casualty and they are still part of the circuit.

3. Repeat, so that two mouth-to-mouth breaths are given.
4. Lean over the casualty, press down on their breastbone, then release. This is a chest compression.
5. Do 15 chest compressions at a rate of roughly one compression per second.
6. Give two more mouth-to-mouth breaths then another 15 compressions.
7. Give mouth-to-mouth breaths every 30 compressions until help arrives.
8. Check the casualty's breathing and pulse at least every 2 minutes.

REMEMBER

There are variations of CPR and guidelines are updated regularly. Check the HSE website for current guidelines.

Recording accidents and near misses

RIDDOR requires employers to:

- report any injuries, diseases or dangerous occurrences to the HSE
- maintain an accident book.

Following an accident, you may need to complete an accident report form. This is usually completed by the person who was injured.

On the accident report form you need to note down:

- the casualty's personal details (e.g. name, age, occupation)
- the name of the person filling in the report form
- the details of the accident.

In addition, the person reporting the accident will need to sign the form.

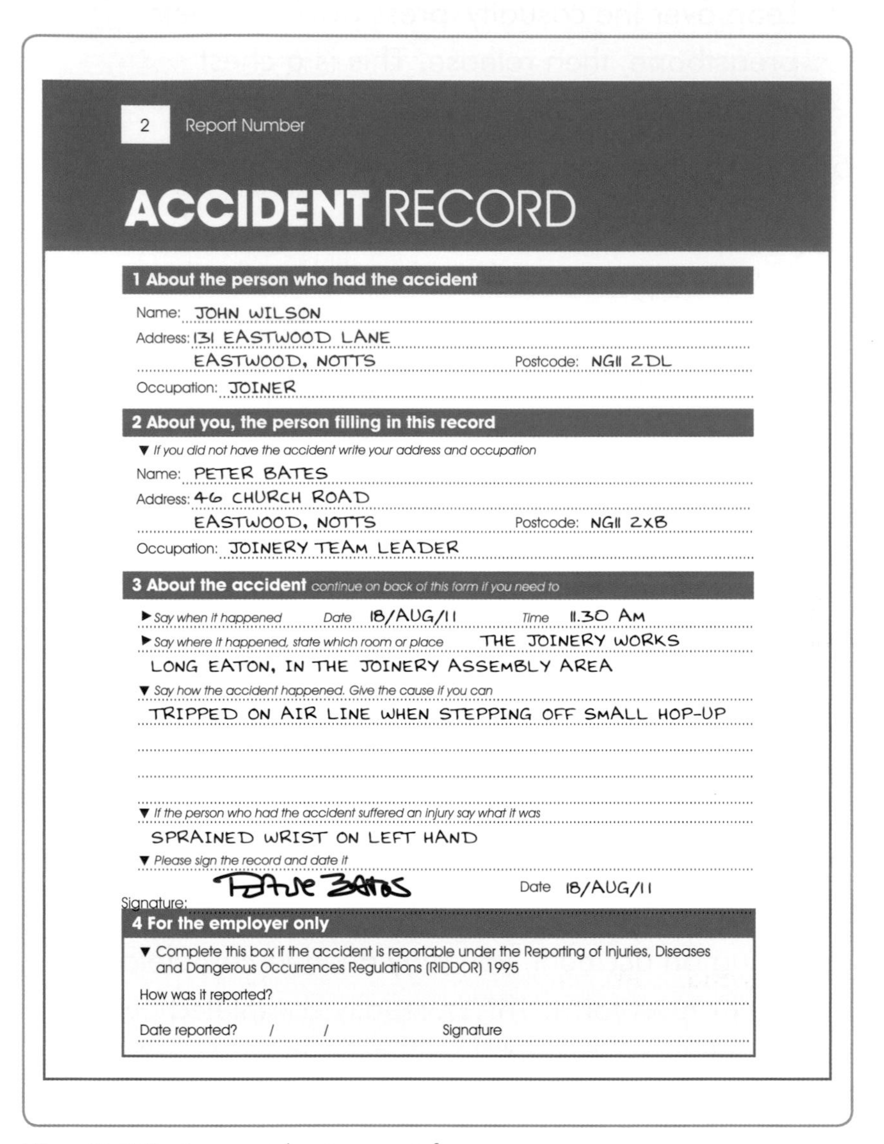

2 Report Number

ACCIDENT RECORD

1 About the person who had the accident

Name: JOHN WILSON
Address: 131 EASTWOOD LANE
EASTWOOD, NOTTS Postcode: NG11 2DL
Occupation: JOINER

2 About you, the person filling in this record

▼ *If you did not have the accident write your address and occupation*
Name: PETER BATES
Address: 46 CHURCH ROAD
EASTWOOD, NOTTS Postcode: NG11 2XB
Occupation: JOINERY TEAM LEADER

3 About the accident *continue on back of this form if you need to*

► *Say when it happened* Date 18/AUG/11 Time 11.30 AM
► *Say where it happened, state which room or place* THE JOINERY WORKS
LONG EATON, IN THE JOINERY ASSEMBLY AREA
▼ *Say how the accident happened. Give the cause if you can*
TRIPPED ON AIR LINE WHEN STEPPING OFF SMALL HOP-UP
▼ *If the person who had the accident suffered an injury say what it was*
SPRAINED WRIST ON LEFT HAND
▼ *Please sign the record and date it*
Signature: Peter Bates Date 18/AUG/11

4 For the employer only

▼ Complete this box if the accident is reportable under the Reporting of Injuries, Diseases and Dangerous Occurrences Regulations (RIDDOR) 1995
How was it reported?
Date reported? / / Signature

Fig. 2.13 An accident report form

ACTIVITY

1. You have a minor cut on your hand. What steps should you take now?
2. One of your work colleagues has fallen off a ladder, dislocating his shoulder and splitting his lip. He has now gone to hospital. What information do you need in order to complete the accident report form?

Electrical safety

It is essential that anyone working with electricity is competent and understands the common dangers.

TOOLBOX TALK

There is inadequate over-current protection in a fifth of all private dwellings.

Common electrical dangers

You are likely to encounter a number of potential dangers when working with electricity on construction sites or in private dwellings. The table below outlines the most common dangers.

Identifying electrical dangers

Danger	Identifying the danger
Faulty electrical equipment	Visually inspect for signs of damage. Equipment should be double insulated or incorporate an earth cable.
Damaged or worn cables	Check for signs of wear or damage regularly. This includes checking power tools and any wiring in the property.
Trailing cables	Cables lying on the ground, or worse, stretched too far, can present a tripping hazard. They could also be cut or damaged easily.
L2 Cables and pipework	Always treat services you find as though they are live. This is very important as services can be mistaken for one another. You may have been trained to use a cable and pipe locator that finds cables and metal pipes.
Buried or hidden cables	Make sure you have plans. Alternatively, use a cable and pipe locator, mark the positions, look out for signs of service connection cables or pipes and dig trial holes by hand to confirm positions.
Inadequate over-current protection	Check circuit breakers and fuses are the correct size current rating for the circuit. A qualified electrician may have to identify and label these.

DID YOU KNOW

According to the HSE a quarter of all electrical accidents involve portable electric appliances.

Fig. 2.14 There are many electrical dangers on a construction site

TOOLBOX TALK

Battery-powered tools are safer than mains-operated tools and equipment. Use 110V tools instead of 230V tools if you are using mains or temporary power supplies. If you are using a 230V power tool then it must be protected by a residual current device (RCD).

Safe use of electrical tools on site

Whether you are using electrical tools or equipment on site, you should always:

- check the plug is in good order
- confirm that the fuse is the correct rating
- check the cable (including making sure that it does not present a tripping hazard)
- find out where the mains switch is, in case you need to turn off the power in the event of an emergency
- make sure electrical equipment is repaired by a qualified electrician
- disconnect from the mains power before making adjustments, such as changing a drill bit
- make sure that the electrical equipment has a sticker that displays a recent test date.

Visual inspection of power tools

Visual inspection and testing is a three-stage process:

1. The user should check for potential danger signs, such as a frayed cable or cracked plug.
2. A formal visual inspection should then take place. If this is done correctly, then 90% of all faults can be detected.
3. Combined inspections and portable appliance testing (PAT) should be conducted at regular intervals by a competent person.

Checking for a valid PAT test

All power tools should be checked before use. A PAT programme of maintenance, inspection and testing is necessary. The frequency of inspection and testing will depend on the appliance. Equipment is usually used for a maximum of three months between each test.

Inspection for general condition

Watch out for the following causes of accidents – they would also fail a safety check:

- Damage to the power cable or plug
- Taped joints on the cable
- Wet or rusty tools and equipment
- Weak external casing
- Loose parts or screws
- Signs of overheating
- The incorrect fuse
- Lack of cord grip
- Electrical wires attached to incorrect terminals
- Bare wires.

HAVE A GO...

Choose a power tool or a common piece of equipment that you use regularly. Carry out a visual check following the guidelines. Do you think it would pass a PAT test? If not, why not?

When tools and equipment fail safety checks

If a tool or piece of equipment fails a safety check then it should be labelled as faulty and taken out of use. It should also have the plug removed to prevent it from being used.

TOOLBOX TALK

Many electricians believe that 'locking off' a circuit is unnecessary when warning notices have been displayed on isolators. However, it is good practice and can help to prevent accidents.

L2 Safe isolation procedure (cutting electrical power)

When preparing to work on an electrical circuit, make sure the circuit is broken before you begin. A 'dead' circuit will not cause you, or anybody else, harm.

These steps must be followed:

1. Switch off – ensure the supply to the circuit is switched off by disconnecting the supply cables or using an isolating switch.
2. Isolate – disconnect the power cables or use an isolating switch.
3. Warn others – to avoid someone reconnecting the circuit, place warning signs at the isolation point.
4. Lock off – this step physically prevents others from reconnecting the circuit.
5. Testing – this involves three steps:
   a Test a voltmeter on a known good source (a live circuit) so you know it is working properly.
   b Check that the circuit to be worked on is dead.
   c Recheck your voltmeter on the known live source, to prove that it is still working properly.

It is important to make sure that the correct point of isolation is identified. Isolation can be next to a local isolation device, such as a plug or socket, or a circuit breaker or fuse.

The isolation should be locked off using a unique key or combination. This will prevent access to a main isolator until the work has been completed. Alternatively, the handle can be made detachable in the OFF position so that it can be physically removed once the circuit is switched off.

L2 Temporary continuity bonding

Temporary continuity bonds are crocodile clips connected by a cable. They are used when disconnecting or reconnecting pipework where a spark could cause a hazard. This is particularly true for gas pipework.

Fig. 2.15 Continuity bonds

Safe use of access equipment

There may be situations where you need to work at height in order to reach something. These situations can include:

- roof work
- loft work
- working on high ceilings.

Access equipment includes all types of ladder, scaffold and platform. You must always use a working platform that is safe. Sometimes a simple step ladder will be sufficient, but at other times you may have to use a tower scaffold.

Generally, ladders are fine for small, quick jobs. However, for larger, longer jobs a more permanent piece of access equipment will be necessary.

TOOLBOX TALK

If you are working from a step ladder or leaning ladder, avoid the temptation to stretch too far.

Types of access equipment and safety checks

The following table outlines the common types of equipment used to work at heights, along with the basic safety checks necessary.

Access equipment – features and safety checks

Equipment	Main features	Safety checks
Step ladder	Ideal for confined spaces Four legs give stability	■ Knee should remain below top of steps ■ Check hinges ■ Position only to face work
Leaning ladder	Ideal for basic access, short-term work Made from aluminium, fibreglass or wood	■ Check rungs, tie rods, repairs ■ Ensure it is placed on firm, level ground ■ Angle should be no greater than 75°
L2 Mobile mini towers or scaffolds	These are usually aluminium and foldable, with lockable wheels	■ Ensure the ground is even and the wheels are locked ■ Never move the platform while it has tools, equipment or people on it
Roof ladders and crawling boards	The roof ladder allows access while, crawling boards provide a safe passage over tiles	■ The ladder needs to be long enough and supported ■ One person should access the roof and another should pass equipment
Mobile tower scaffolds	These larger versions of mini towers usually have edge protection	■ Ensure the ground is even and the wheels are locked ■ Never move the platform while it has tools, equipment or people on it ■ Base width-to-height ratio should be no greater than 1 : 3
Fixed scaffolds and edge protection	Scaffolds fitted and sized to the specific job, with edge protection and guard rails	■ There needs to be sufficient braces, guard rails and scaffold boards ■ The tubes should be level ■ There should be proper access using a ladder
Mobile elevated work platforms (known as scissor lifts or cherry pickers)	Can be used indoors or outside Work can be reached quickly and easily Has guard rails and toe boards	■ Use guard rails and toe boards ■ Care needs to be taken to avoid overhead hazards

Assembling and using access equipment

- Step ladders should always rest firmly on the ground. Only use the top step if the ladder is part of a platform.
- Do not rest leaning ladders against fragile surfaces. Always use both hands to climb. It is best if the ladder is steadied by someone at the foot of the ladder.
- A roof ladder is positioned by turning it on its wheels and pushing it up the roof. It then hooks over the ridge tiles. Ensure that the access ladder to the roof is directly beside the roof ladder.
- A mobile tower scaffold is put together by slotting sections together until the required height is reached. The working platform needs to have a suitable edge protection. Always push from the bottom of the base and not from the top to move it, otherwise it may lean or topple over.

TOOLBOX TALK

When carrying out electrical work, never use an aluminium ladder as it will conduct electricity if it comes into contact with a live wire.

Fig. 2.16 A tower scaffold

ACTIVITY

A client wants their rear guttering cleared and the back of their house repainted. Suggest suitable access equipment for this job, bearing in mind the following:

- There is limited access to the rear of the property via a 1.5-metre-wide passageway.
- The rear of the property opens straight onto lawn.
- Overhead telephone cables are at gutter height.

Working safely with heat-producing equipment

It is important to know how to work safely with heat-producing equipment used for pipe and sheet joining. You need to be able to identify the gases, know how to transport them, store them, check them for safety, and assemble them.

Types of gases

All gas cylinders in the UK are colour-coded.

- Oxygen cylinders are black with a white or grey shoulder.
- Acetylene cylinders are maroon.

The gases are stored in the cylinders at high pressure.

REMEMBER

Acetylene is a flammable gas, as is propane.

Oxyacetylene torches are used for welding metal. There are other oxy-fuel systems, which include LPG (liquid petroleum gas) cylinders, such as:

- propane–oxygen (propane is usually a red bottle)
- butane–oxygen (butane is usually a blue bottle).

L2 Uses in the industry

The uses of gases in the industry include:

- sheet lead welding (known as 'burning')
- steel pipe brazing
- copper tube soldering.

L2

Lead burning and steel pipe brazing use oxyacetylene torches, which burn at a higher temperature. Copper tube soldering requires less heat, so an LPG like propane burning in air is used.

Gasses and their use in the industry

Type	Set up	Burning temperature	Examples of uses
Oxyacetylene	Two tanks, two hoses – one for oxygen and one for acetylene fuel	Around 3500 °C	Cutting steel, soldering and brazing, and welding
Propane–oxygen	Two tanks, two hoses – one for oxygen and one for propane fuel	Around 2500 °C	Cutting when an injector-style torch is used Does not weld steel
Propane–air	One tank, one hose	Around 2000 °C	Soldering (e.g. jointing copper pipe for plumbing)

Transporting and storing bottled gases

Special safety measures are needed when working with bottled gases as they can be hazardous if they are not handled properly. You must remember these safety points:

- Always keep the cylinders upright and restrain them to stop them from falling over.
- Never drop, roll or drag the cylinders.
- When transporting cylinders, fit protective valve caps.
- Disconnect regulators and hoses whenever practicable.
- Never let the cylinders project beyond the sides of a vehicle.
- Always make sure that the cylinders are clearly marked.
- When removing cylinders from storage, always make sure that the oldest cylinders are used first.

DID YOU KNOW

Oxyacetylene torches have interchangeable heads, depending on the type of material that is being welded.

Fig. 2.17 Gas cylinders

L2

- Keep all cylinders in a cool, dry, clean, well-ventilated space. They must not be at risk of physical damage and must be on a flat surface.
- Never store cylinders so they stand or lie in water.

TRADE TIP

Do not try to repair hoses and fittings.

ACTIVITY

You are working in an industrial unit and have just run out of gas for the oxyacetylene equipment. The owner of the unit tells you there are spare cylinders in the store room. When you look, all of the cylinders are rusty and it is difficult to work out the original colours of the cylinders. What should you do?

Heat-producing equipment safety

Before you carry out any welding, it is important to check the condition and operation of the equipment. A basic oxyacetylene torch has a number of key pieces of equipment, as described in the table opposite.

Oxyacetylene torch equipment and basic safety checks

Equipment	Safety check L2
Hoses	These should be positioned between the torch and the gas regulator They should be colour-coded: acetylene should be red; oxygen should be blue Each hose has a different thread direction: the oxygen hose has a right-hand thread; the acetylene hose has a left-hand thread
Flashback arresters	Also known as flame traps they should be fitted to both gas lines to prevent flashback flame from reaching the regulators
Control valves	These control the gas pressure and reduce the high pressure of the bottle-stored gas to the working pressure of the torch
Gauges	There should be two – a high-pressure gauge for the gas in the cylinder and a low-pressure gauge for the pressure of gas that is being fed to the torch
Blowpipes	These range from light-duty to heavy-duty The nozzle should have a diameter according to the thickness of the material being worked
Direct connecting combined units	These are complete units with a twin gauge regulator They are often turbo-fan cooled

Safe assembly and use of gas heating equipment

It is important to ensure that safety procedures are followed when assembling and using gas heating equipment. The first thing to remember is that you should be using PPE, which could include:

- work clothing, including footwear
- eye protection
- ear protection
- hand protection
- respiratory equipment.

You should then go through the following steps:

1. Check the equipment – is the gas flow turned off? Are the cylinders secure and upright? The gauges should read 0.
2. Purge the system – close the main cylinder valve, pick up the torch and check the hose attachments, then turn the oxygen regulator clockwise and open the oxygen valve. This will purge the gas in the system. Repeat with the acetylene cylinder.
3. Torch handle – examine all connections, making sure the one marked OX is attached to the oxygen hose and the one marked AC is attached to the acetylene hose.
4. Connect hoses – the oxygen connector is right-hand threaded and the acetylene connector is left-hand threaded.
5. Install tip – select the tip size suitable for your heating task and screw it on.
6. Adjust pressure – close the oxygen valve clockwise and then the acetylene valve clockwise to adjust the pressure of the gas flow.
7. Turn on the gas – turn the oxygen to 10 psi and the acetylene to 5 psi.
8. Check work area for flammable materials.
9. Put on tinted goggles or face mask.
10. Ignite the torch.
11. Adjust the flame.

When finished, extinguish the oxyacetylene flame with the following procedure:

1. Turn off the torch's acetylene valve first.
2. Then turn off the torch's oxygen valve.
3. Wait for the flame to go out.
4. Close the cylinder valves.
5. Open and close the torch's oxygen valve to release the pressure.
6. Open and close the torch's acetylene valve.

Safety checks

Before you ignite gas heating equipment, check it for leaks. To do this, spray leak detector on the joints and along the length of the hoses. If you see bubbles, this indicates a leakage of gas.

If you discover a leak, tighten any leaking connections or replace any faulty hose. Then check again to confirm no leakages are present.

When using the heating equipment, make sure that the cylinders and hose are away from what is being heated. The gas bottles should be away from the immediate work area, but within reach to turn off in an emergency situation.

When using gas heating equipment you should also:

- never use oil or grease with oxygen equipment as it acts explosively
- purge hoses before lighting
- only use in well-ventilated areas
- never use an oxygen or acetylene welding torch to remove dust from clothing.

Check gas heating equipment for leaks if at any time you suspect one. Do not use soapy water to check for leaks as it can react with the acetylene.

The fire triangle

Fig. 2.18 The fire triangle

You need three ingredients before anything will burn (combust). These are:

- oxygen
- heat
- fuel.

The fuel can be anything that burns, such as wood, paper, or flammable liquids or gases. Oxygen is in the air around us, so all that is needed to start a fire is sufficient heat.

The fire triangle is used to represent these three elements visually. By removing one of the three elements the fire can be prevented or extinguished.

Dangers and preventing fires

Never allow acetylene to come into contact with copper as it will explode.

You need to be very careful when using gas cylinders. The main hazards are gas cylinder explosions and fires from the escape of flammable gases.

Remember these safety tips when using gas heating equipment:

- Always ensure that the cylinders are securely stored in an upright position, away from sources of ignition or flammable materials.
- Normal work clothing will not provide sufficient protection, so always wear proper welding goggles and use a heat shield behind whatever you are heating up.
- Never point the lighted flame towards another person or flammable material.

- Always light the torch with a striker and not a lighter or a match.
- Have a suitable fire extinguisher close at hand.
- If you have been working on a piece of metal, write 'hot' on it with a piece of chalk.
- If there is a flashback or fire, plunge the nozzle of the torch into water, but leave the oxygen running to prevent water from entering the blowpipe.

TRADE TIP

If the cylinder falls over and the main valve breaks, the cylinder will be very dangerous.

Fighting localised fires

Extinguishers can be effective when tackling small localised fires. However, you must use the correct type of extinguisher. For example, putting water on an oil fire could make it explode.

When using an extinguisher it is important to remember the following safety points:

- Only use an extinguisher at the early stages of a fire, when it is small.
- The instructions for use appear on the extinguisher.
- If you do choose to fight the fire because it is small enough and you are sure you know what is burning, position yourself between the fire and the exit, so that if it doesn't work you can still get out.

REMEMBER

Do not attempt to put out a fire if you're not sure you can handle it or if it's spreading from its starting point.

The different types of extinguisher and their uses are shown in the table overleaf.

Fire blankets can be used to extinguish any type of fire described above. They can also be used when clothing is alight since it does not pose a risk to skin or breathing, as some extinguishers do.

Fire extinguishers and their uses

Type of fire risk	Fire class Symbol	White label Water	Cream label Foam	Black label Carbon dioxide	Blue label Dry powder	Yellow label Wet chemical
A – Solid (e.g. wood or paper)	A	✓	✓	✗	✓	✓
B – Liquid (e.g. petrol)	B	✗	✓	✓	✓	✗
C – Gas (e.g. propane)	C	✗	✗	✓	✓	✗
D – Metal (e.g. aluminium)	D METAL	✗	✗	✗	✓	✗
E – Electrical (i.e. any electrical equipment)	E	✗	✗	✓	✓	✗
F – Cooking oil (e.g. a chip pan)	F	✗	✗	✗	✗	✓

DID YOU KNOW

You can find out more about the different types of extinguisher and their uses by going to: www.firesafe.org.uk/types-use-and-colours-of-portable-fire-extinguishers/

There are some differences you should be aware of when using different types of extinguisher:

- CO_2 extinguishers – do not touch the nozzle; simply operate by holding the handle. This is because the nozzle gets extremely cold when ejecting the CO_2, as does the canister. Fires put out with a CO_2 extinguisher may reignite, and you will need to ventilate the room after use.
- Powder extinguishers can be used on lots of kinds of fire, but can seriously reduce visibility by throwing powder into the air as well as on the fire.

L2 Excavations and confined spaces

Any excavation work needs to be properly planned, managed and supervised in order to prevent accidents. It may be necessary to work in excavations in order to access pipework and services or to install them.

Fig. 2.19 Excavation begins

Safe access into the excavation

Good ladder access or other safe ways of getting in and out of the excavation should always be provided.

Trench support systems

It is also important to make sure that the excavation is protected from collapse. The sides and ends of trenches should be supported with timber, sheeting or other supporting material.

Even shallow trenches are dangerous. If you need to bend or kneel in them then they should be supported.

REMEMBER

You should never go into an unsupported excavation or work ahead of the support.

Preventing people or equipment from falling into excavations

Precautions need to be taken to stop people or objects from falling into the excavation:

- If the excavation is more than 2 m deep, use guard rails and toe boards.
- To keep vehicles away, put up brightly painted barriers.

Always wear a hard hat when working in excavations.

L2

- If vehicles have to tip material into the excavation, use stop blocks. Also, the sides of the excavation may need extra support.
- Spoil and other materials should not be stored close to the excavation, as they may fall in.
- Toe boards can be used to protect the edges of the excavation.
- The excavation should be fenced off.
- If the public can get on to the site out of hours, the excavation should either be backfilled or securely covered.

Working in confined spaces

KEY TERMS

Confined space: A working area that is substantially enclosed, in which accident or injury could occur.

You may sometimes need to work in a **confined space**, for example when dealing with:

- drainage systems
- plant rooms
- duct rooms
- tanks, cylinders, boilers or cisterns
- suspended timber floors
- roof spaces.

The Confined Spaces Regulations (1997) suggest that it is best to avoid confined spaces. If this is not possible, then safe work systems are necessary.

The main dangers of working in a confined space are:

- a lack of oxygen – particularly where there is a reaction between soil and oxygen, or inside tanks and vessels where rust has formed

L2

- poisonous gases or vapours, which can build up in sewers, manholes, pits and trenches
- fire and explosions, which can ignite from flammable vapours or a build up of oxygen
- dust present under timber floors or in roof spaces, which would restrict breathing
- hot conditions leading to overheating, particularly in places exposed to the sun or where machinery is in operation.

Safe working in confined spaces

If it is impossible to avoid working in a confined space, you need to carry out a risk assessment in order to establish a safe system for working. This may include the following safety points:

- A supervisor should be given overall responsibility.
- Only experienced workers should be used.
- Mechanical and electrical isolation of equipment.
- The space should be cleaned before entering.
- The entrance size should be checked.
- Additional ventilation should be considered and the air tested.
- Breathing apparatus should be provided.
- Emergency arrangements should be made available.
- There should be communication between those inside and those outside.

CHECK YOUR KNOWLEDGE

LEVEL 1

1. What are the two main pieces of health and safety legislation?
2. Give an example of a mandatory sign.
3. What hazardous substance might be found in Artex?
4. Give two examples of common hearing protection.
5. How would you check that it is safe to move a load?
6. What do CPR and ABC stand for?
7. How many people should be used when a ladder is necessary for work and why?
8. What would you do in order to immediately extinguish the flame of a welding torch?

CHECK YOUR KNOWLEDGE

LEVEL 2

1. What is the HSE and what is its purpose?
2. Describe the five main types of safety notice and give an example of each.
3. What is the common name for chrysotile and where might it be found?
4. What is a PAT test and how is it carried out?
5. How would you isolate a piece of equipment in an emergency?
6. List the main parts of a welding kit and their purpose.
7. How should a roof ladder be secured?
8. State three potential dangers of working in a confined space.

3 Environmental protection measures

This chapter covers the learning outcomes for:

Understand fundamental environmental protection measures within building services engineering

L2 Understand how to apply environmental protection measures within building services engineering

City & Guilds unit number 102 and L2 202; EAL unit code QACC1/02 and L2 QBSE2/03; ABC unit code A03 and L2 A07.

Building services engineering has a vital role to play in the protection of the environment, waste reduction and conserving energy. There is also the requirement for correct waste disposal, both of recyclable materials and any material that is potentially hazardous. You will explore these responsibilities further in this chapter.

There is an increasing amount of legislation on energy conservation that is relevant to building services engineering. There is also a wide variety of new technologies, which can be incorporated into building projects, that aim to reduce waste and conserve energy. In this chapter you will look at how these new, green technologies can help to reduce carbon emissions, conserve natural resources and encourage recycling.

IN THIS CHAPTER YOU WILL LEARN ABOUT:

- energy conservation legislation
- application of energy sources
- energy conservation and commissioning
- reducing waste and conserving energy
- safe disposal of materials
- conserving and reducing the wastage of water.

L2 Energy conservation legislation

Carbon is present in all fossil fuels, such as coal or natural gas. Burning fossil fuels releases carbon dioxide, which is a greenhouse gas linked to climate change.

Energy conservation aims to reduce the amount of carbon dioxide in the atmosphere. The idea is to do this by making buildings better insulated and, at the same time, make heating appliances more efficient. It also means attempting to generate energy using renewable and/or low or zero carbon methods.

DID YOU KNOW

New minimum requirements came into force in October 2010. There were revisions to Parts G, F, J and L of the Building Regulations.

Building Regulations (2010)

In terms of energy conservation, the most important UK law is the Building Regulations (2010), particularly Part L.

The Building Regulations:

- list the minimum efficiency requirements
- provide guidance on compliance, the main testing methods, installation and control
- cover both new dwellings and existing dwellings.

KEY TERMS

Renewable source: An energy source that is constantly replaced and will never run out, such as water, wind and solar energy.

A key part of the regulations is the Standard Assessment Procedure (SAP), which measures or estimates the energy efficiency performance of buildings.

Local planning authorities also now require that all new developments generate at least 10% of their energy from **renewable sources**. This means that each new project has to be assessed one at a time.

L2 Water Regulations

In the UK, the Water Supply (Water Fittings) Regulations (1999) aim to prevent water contamination as well as the waste and misuse of water.

With regard to the building services, this legislation has led to a number of new developments and existing dwellings incorporating waste water reuse and the recycling of rainwater.

Energy conservation – who is responsible?

By law, each local authority is required to reduce carbon dioxide emissions and to encourage the conservation of energy. This means that everyone has a responsibility in some way to conserve energy.

- Clients, along with building designers, are required to include energy-efficient technology in the build.
- Contractors and sub-contractors have to follow these design guidelines. They also need to play a role in conserving energy and resources when actually working on site.
- Suppliers of products are required by law to provide information on energy consumption.

In addition, new energy-efficiency schemes and Building Regulations cover the energy performance of buildings. Each new build is required to have an Energy Performance Certificate. This rates a building's energy efficiency from A (very efficient) to G (least efficient).

Some building designers have also begun to adopt other voluntary ways of attempting to protect the

L2

environment. These include BREEAM, which is an environmental assessment method, and the Code for Sustainable Homes, which is a certification of **sustainability** for new builds.

KEY TERMS

Sustainability: In terms of building services engineering, this is about reducing a building's environmental impact over its lifetime.

TRADE TIP

You can find out more about the Energy Saving Trust at: www.energysavingtrust.org.uk

Fig. 3.1 The Energy Saving Trust encourages builders to use less wasteful building techniques and more energy-efficient construction

Application of energy sources

Types of energy

When we look at energy sources, we consider their environmental impact in terms of how much carbon dioxide they release. Accordingly, energy sources can be split into three different groups:

- High carbon – those that release a lot of carbon dioxide
- Low carbon – those that release some carbon dioxide
- Zero carbon – those that do not release any carbon dioxide

Some examples of high, low and zero carbon energy sources are given in the tables below.

High carbon energy source	Description
Natural gas or LPG	Piped natural gas or liquid petroleum gas (LPG) stored in bottles
Fuel oils	Domestic fuel oil, such as diesel
Solid fuels	Coal, coke and peat
Electricity	Generated from non-renewable sources, such as coal-fired power stations

Low carbon energy source	Description
Solar thermal	Panels used to capture energy from the sun to heat water
Solid fuel	Biomass such as logs, wood chips and pellets
Hydrogen fuel cells	Converts chemical energy into electrical energy
Heat pumps	Devices that convert low temperature heat into higher temperature heat
Combined heat and power (CHP)	Generates electricity as well as heat for water and space heating
Combined cooling, heat and power (CCHP)	A variation on CHP that also provides a basic air conditioning system

Zero carbon energy	Description
Electricity/wind	Uses natural wind resources to generate electrical energy
Electricity/tidal	Uses wave power to generate electrical energy
Hydroelectric	Uses the natural flow of rivers and streams to generate electrical energy
Solar photovoltaic	Uses solar cells to convert light energy from the sun into electricity

It is important to try to conserve non-renewable energy so that there will be sufficient fuel for the future. The

ACTIVITY

1. Think about where you live. Does the dwelling aim to reduce demand and improve efficiency?
2. How many of the basic environmental technology systems are part of the dwelling? Use Figure 3.2 as a guide.

idea is that the fuel should last as long as is necessary to completely replace it with renewable sources, such as wind or solar energy.

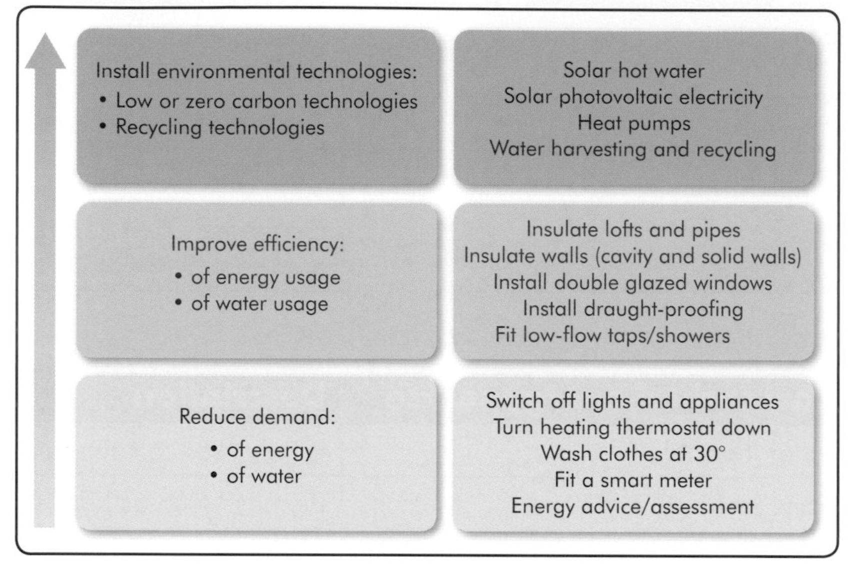

Fig. 3.2 Working towards reducing carbon emissions

Basic installation principles

Each environmental energy installation has its own particular operating principles. Many of them use energy from different sources.

Solar thermal systems

At the heart of this system is the solar collector, which is often referred to as a solar panel. The idea is that the collector absorbs the sun's energy, which is then converted into heat. This heat is then applied to the system's heat transfer fluid.

The system uses a differential temperature controller (DTC) that controls the system's circulating pump when solar energy is available and there is a demand for water to be heated.

In the UK, due to the lack of guaranteed solar energy, solar thermal hot water systems often have an auxiliary heat source, such as an immersion heater.

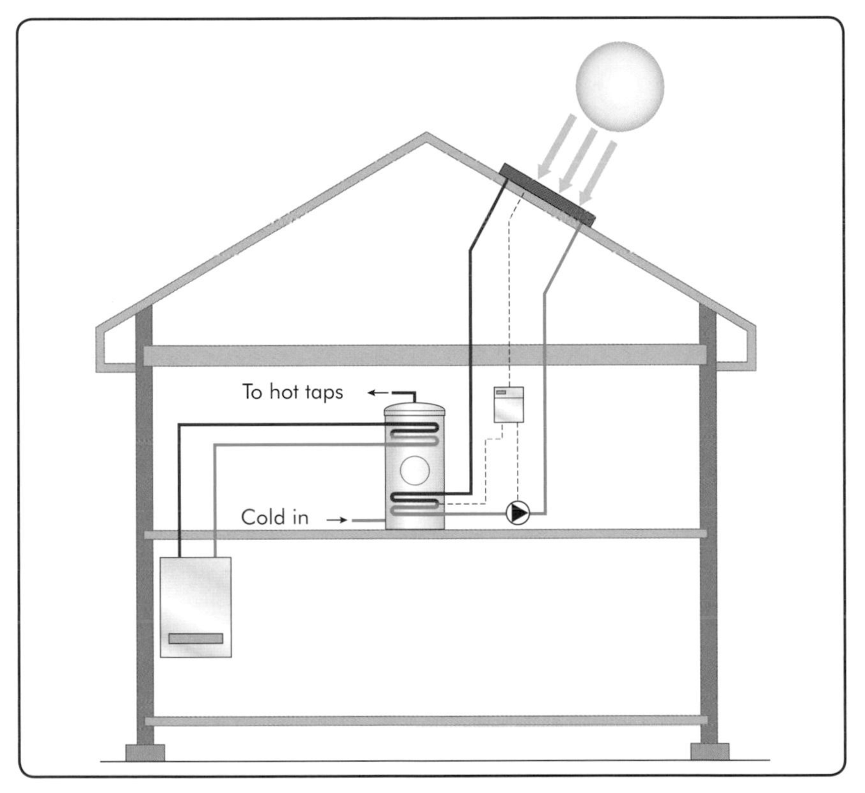

Fig. 3.3 Solar thermal hot water system

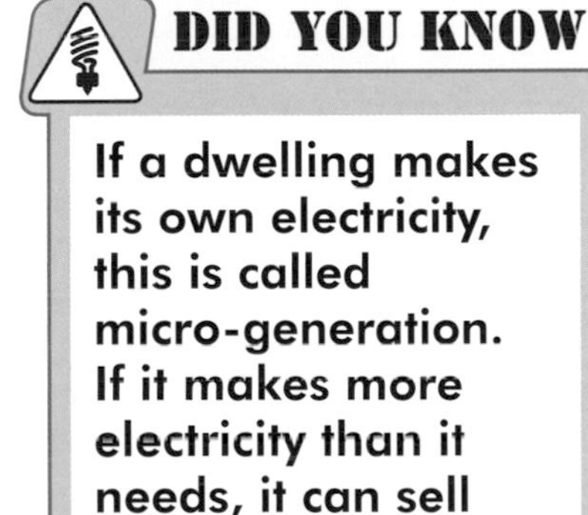

DID YOU KNOW

If a dwelling makes its own electricity, this is called micro-generation. If it makes more electricity than it needs, it can sell that electricity to the National Grid.

L2 Biomass (solid fuel)

Biomass stoves burn either pellets or logs. Some have integrated hoppers that transfer pellets to the burner. Biomass boilers are available for pellets, woodchips or logs. Most of them have automated systems to clean the heat exchanger surfaces. They can provide heat for domestic hot water and space heating.

KEY TERMS

Heat sink: A heat exchanger that transfers heat from one source into a fluid, such as in refrigeration, air conditioning or the radiator in a car.

Heat pumps

Heat pumps convert low temperature heat from air, ground or water sources to higher temperature heat. They can be used in ducted air or piped water **heat sink** systems.

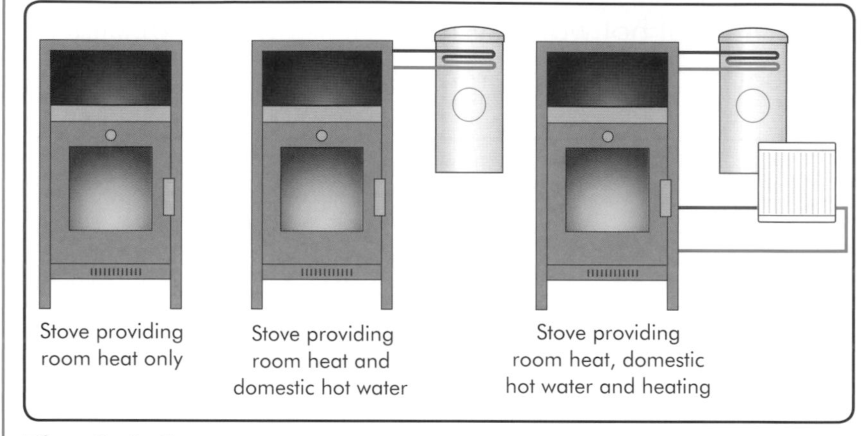

Fig. 3.4 Biomass stoves output options

There are a variety of different arrangements for each of the three main systems:

- Air source pumps operate at temperatures as low as −20 °C. They have units that receive incoming air through an inlet duct.
- Ground source pumps operate on **geothermal** ground heat. They use a sealed circuit collector loop, which is buried either vertically or horizontally underground.
- Water source systems can be used where there is a suitable water source, such as a pond or lake.

KEY TERMS

Geothermal: Relating to the internal heat energy of the earth.

The heat pump system's efficiency relies on the temperature difference between the heat source and the heat sink. Special tank hot water cylinders are part of the system, giving a large surface-to-surface contact between the heating circuit water and the stored domestic hot water.

L2

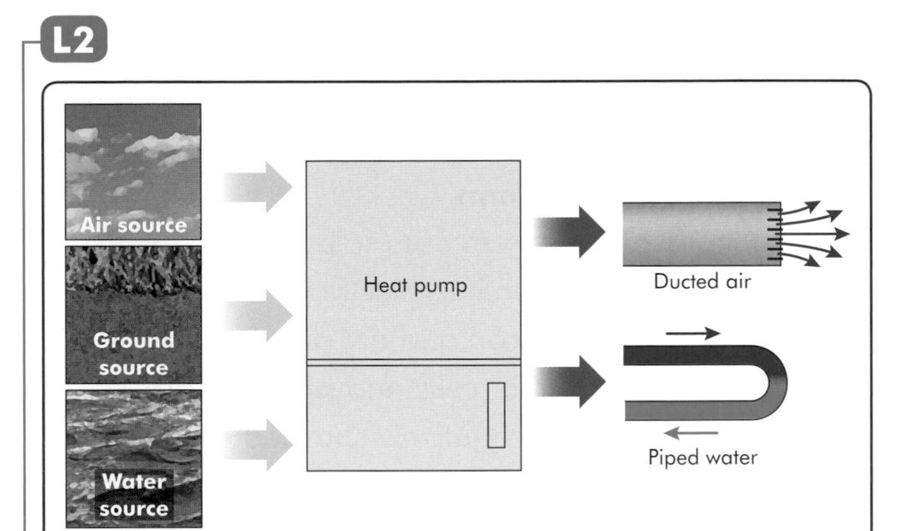

Fig. 3.5 Heat pump input and output options

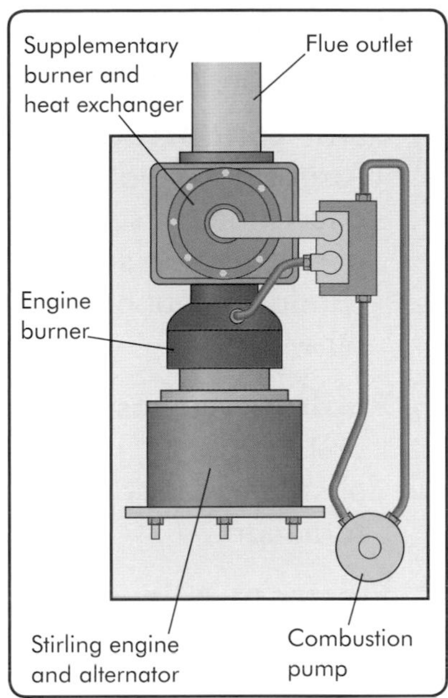

Fig. 3.6 Example of a micro combined heat and power (MCHP) unit

Combined heat and power (CHP) and combined cooling, heat and power (CCHP) units

These are similar to heating system boilers, but they generate electricity as well as heat for hot water or space heating (or cooling). The heart of the system is an engine or gas turbine. The gas burner provides heat to the engine when there is a demand for heat. Electricity is generated along with sufficient energy to heat water and to provide space heating.

CCHP systems also incorporate the facility to cool spaces when necessary.

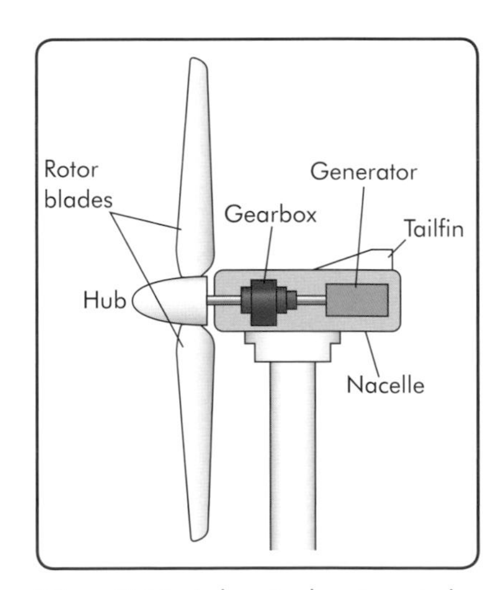

Fig. 3.7 A basic horizontal-axis wind turbine

Wind turbines

Freestanding or building-mounted wind turbines capture the energy from wind to generate electrical energy. The wind passes across rotor blades of a turbine, which causes the hub to turn. The hub is connected by a shaft to a gearbox. This increases the

KEY TERMS

Semi-conductor: A substance that allows the passage of electricity in some conditions but not in others.

Photon: The basic unit of light or electromagnetic radiation.

Electron: A tiny particle that is part of an atom.

Atom: The smallest unit of an element, made up of protons, neutrons and electrons.

Direct current (DC): An electric current that flows in one direction only.

Alternating current (AC): An electric current where the direction of current reverses at regular intervals.

speed of rotation. A high-speed shaft is then connected to a generator that produces the electricity.

Solar photovoltaic systems

A solar photovoltaic system uses solar cells to convert light energy from the sun into electricity. The solar cells are usually made of silicon and are **semi-conductors**. The sunlight hits the solar cells and **photons** are absorbed. This causes negatively charged **electrons** in the cell to detach from their **atoms** and flow through the cell to create electricity. The electricity is **direct current** (DC). The DC current is then converted by an inverter to **alternating current** (AC), which is the type of current used for mains electricity.

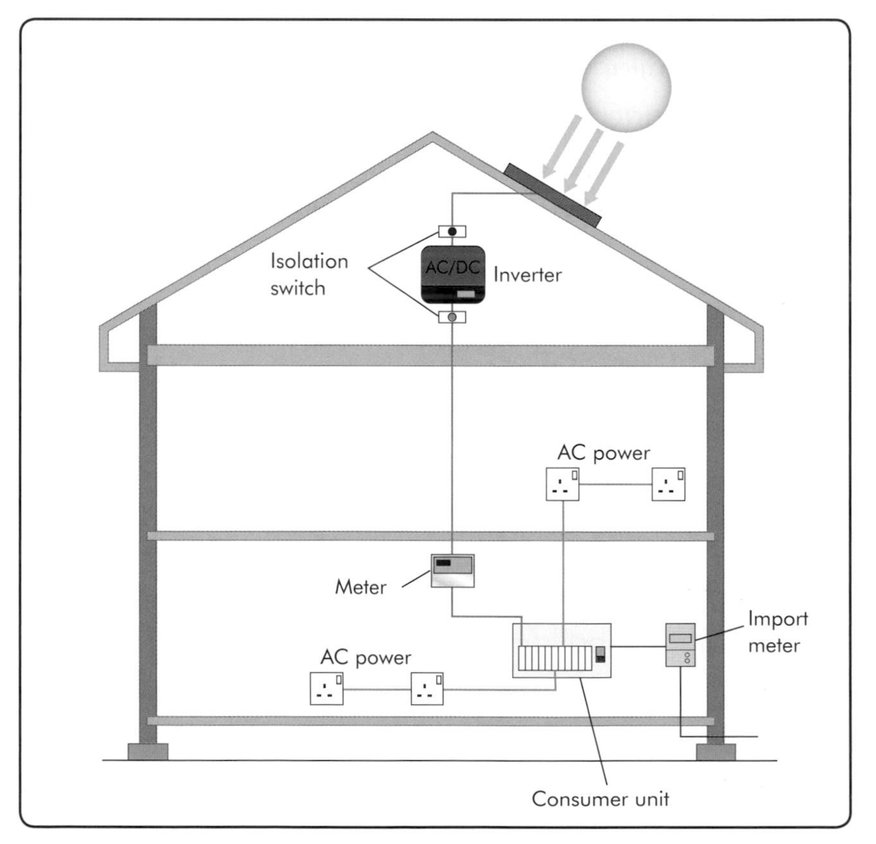

Fig. 3.8 A basic solar photovoltaic system

Guidance and advice on energy-saving and conservation

There are an increasing number of public and private organisations that can provide guidance and advice on energy-saving and conservation techniques. A good place to start is the website of the Energy Saving Trust (**www.energysavingtrust.org.uk**) as it provides links to several other organisations. They also have a network of advice centres across the UK.

Other sources of advice:

- Carbon Trust (**www.carbontrust.co.uk**) provides help in cutting carbon and saving energy.
- Local councils or authorities all have energy and climate change advisers or departments.
- The Department of Energy and Climate Change (**www.decc.gov.uk**) is responsible for government policy on energy and climate change.
- Local electricity suppliers.

L2 Energy rating tables and components

Energy rating tables are used to measure the overall efficiency of a dwelling, with rating A being the most energy efficient and rating G the least energy efficient.

Alongside this, an environmental impact rating (see Fig. 3.10) measures the dwelling's impact on the environment in terms of how much carbon dioxide it produces. Again, rating A is the highest, showing it has the least impact on the environment, and rating G is the lowest.

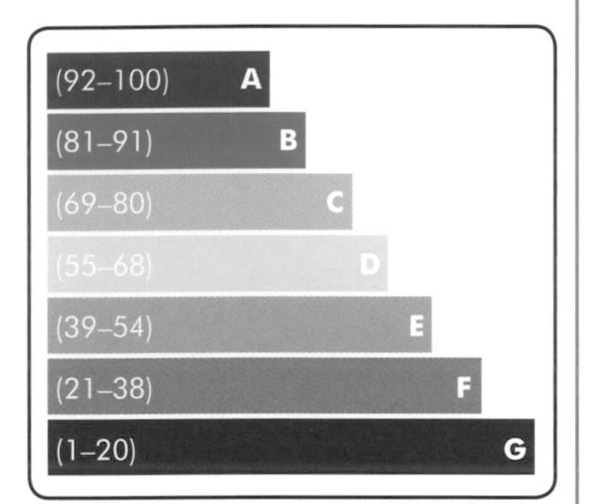

Fig. 3.9 SAP energy efficiency rating table. The ranges in brackets show the percentage of energy efficiency for each banding

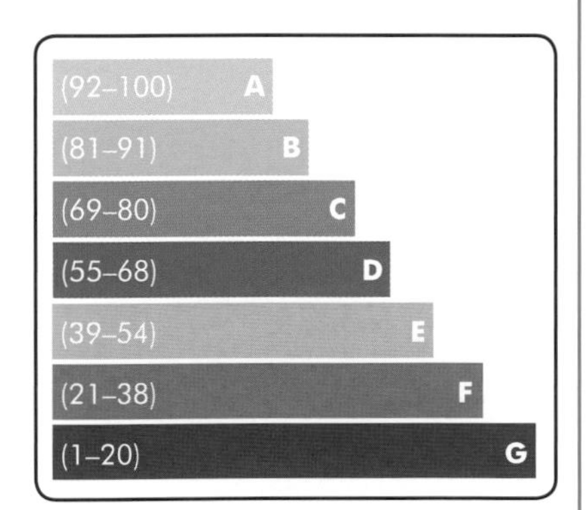

Fig. 3.10 SAP environmental impact rating table

L2 A Standard Assessment Procedure (SAP) is used to place the dwelling on the energy rating table. This will take into account:

- the date of construction, the type of construction and the location
- the heating system
- insulation (including cavity wall)
- double glazing.

The ratings are used by local authorities and other groups to assess the energy efficiency of new and old housing.

Information on alternative energy sources

There are various sources of funding, guidance and advice on energy saving and conservation techniques including the following:

- Energy Share (**www.energyshare.com**) encourages ways to set up renewable energy projects.
- The Renewable Energy Centre (**www.therenewableenergycentre.co.uk**) provides news and links to designers, suppliers, trade associations and government departments.
- Energy Saving Trust (**www.energysavingtrust.org.uk**) provides information about saving energy and reducing carbon emissions.
- Carbon Trust (**www.carbontrust.co.uk**) aims primarily to help businesses in the public sector save energy and cut carbon emissions.

L2

- BRE Trust (**www.bre.co.uk**) is a research group for the built environment.
- National Energy Foundation (**www.nef.org.uk**) aims to develop and implement energy efficient initiatives.
- Carbon Leapfrog (**www.carbonleapfrog.org**) is a charity that supports carbon reduction projects.
- Energy4All (**www.energy4all.co.uk**) aims to expand the number of renewable energy cooperatives.
- Renewable Energy Association (**www.r-e-a.net**) represents the UK's renewable energy industry.

ACTIVITY

Visit some of the websites listed here and find out who can offer grants or funds to assist in the installation of renewable energy systems in your area.

Energy conservation and commissioning

The role of the commissioning process

When a new environmental energy system is installed in a dwelling or a new build, it is important that the installers carry out a rigorous commissioning process. This is done in order to ensure that the system is working to its full efficiency.

Commissioning normally takes place once the installation has been completed. It is a process in which the installers can check all of the components of the installation before it is handed over to the building owner and signed off. The purpose is to check the system's overall performance.

Commissioning and handing over the system to the customer will ensure:

L2

- the system operates in the most effective way from the very beginning
- it is the start of an ongoing and cost-effective maintenance programme
- the owner is satisfied and that it meets their needs.

Commissioning should prevent minor errors from developing into serious problems.

System handover procedure

One of the most important aspects of commissioning and the handover procedure is to provide the owner with accurate documentation and technical reports. These serve as a benchmark for future system tests. Also, it is unlikely that the owner has a thorough working knowledge of the system, so the handover needs to provide basic training for them to operate and routinely maintain the system.

Reducing waste and conserving energy

Working practices

The expectation within the building services industry is increasingly that working practices conserve energy and protect the environment. Everyone can play a part in this. For example, you can contribute by turning off hosepipes when you have finished using water, or unplugging equipment or power tools that you would usually leave on standby.

DID YOU KNOW

About 18% of the UK's electricity usage is for electric lighting, and about 5% goes towards reducing the heat from lights. (Source: *Digest of UK Energy Statistics 2010***)**

Simple things, such as keeping construction sites neat and orderly, can go a long way to conserving energy and protecting the environment. A good way to remember this is:

- Sort – sort and store items in your work area, eliminate clutter and manage deliveries.
- Set – everything should have its own place and be clearly marked and easy to access. In other words, be neat.
- Shine – clean your work area and it will be easier to see potential problems.
- Standardise – using standardised working practices, you can keep organised, clean and safe.

Reducing material wastage

Reducing waste is all about good working practice. By reducing wastage disposal and recycling materials on site, you will benefit from savings on raw materials and lower transportation costs.

Let's start by looking at ways to reduce waste when buying and storing materials:

- Only order the amount of materials you actually need, to reduce over-ordering and potential waste.
- Arrange regular deliveries so you can reduce storage and material losses.
- Think about using recycled materials, as they may be cheaper.
- Consider if all the packaging is absolutely necessary. Can you reduce the amount of packaging?

TOOLBOX TALK

Small, regular deliveries will help to ensure that materials stored on site are neither damaged nor stolen.

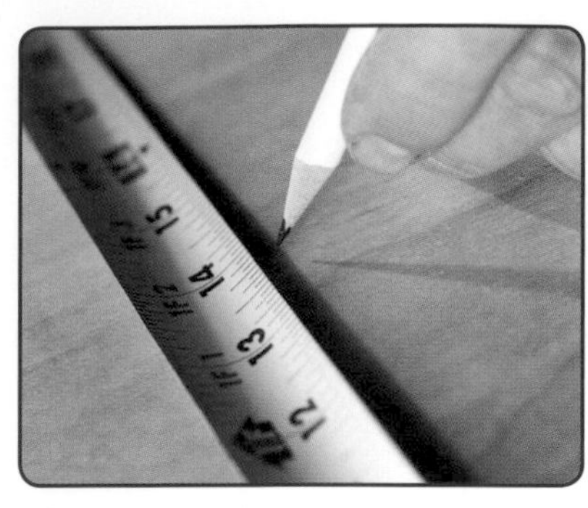

Fig. 3.11 Measuring accurately can reduce wastage

- Reject damaged or incomplete deliveries.
- Make sure that storage areas are safe, secure and weatherproof.
- Store liquids away from drains to prevent pollution.

By planning ahead and accurately measuring and cutting materials, you will be able to reduce wastage.

Safe disposal of materials

L2 Statutory legislation for waste management

By law, all construction sites should be kept in good order and clean. A vital part of this is the proper disposal of waste, which can range from **low-risk waste**, such as metals, plastics, wood and cardboard, to **hazardous waste**, for example asbestos, electrical and electronic equipment and refrigerants.

Waste is anything that is thrown away because it is no longer useful or needed. However, you cannot simply discard it, as some waste can be recycled or reused, while other waste will affect health or the quality of the environment.

Legislation aims not only to prevent waste from going into landfill but also to encourage people to recycle. For example, under the Environmental Protection Act (1990), the building services industry has the following duty of care with regard to waste disposal:

KEY TERMS

Low-risk waste: Waste that is not very dangerous but still needs to be disposed of carefully.

Hazardous waste: Waste that could cause great harm to people or the environment and needs to be disposed of with extreme care.

L2

- All waste for disposal can only be passed over to a licensed operator.
- Waste must be stored safely and securely.
- Waste should not cause environmental pollution.

The main legislation covering the disposal of waste is outlined in the table below.

Fig. 3.12 Low-risk waste

UK waste disposal laws

Legislation	Brief explanation
Environmental Protection Act (1990)	Defines waste and waste offences
Environmental Protection (Duty of Care) Regulations (1991)	Places the responsibility for disposal on the producer of the waste
Hazardous Waste (England and Wales) Regulations (2005)	Defines hazardous waste and regulates the safe management of hazardous waste
Waste Electrical and Electronic Equipment (WEEE) Regulations (2006)	Requires those that produce electrical and electronic waste to pay for its collection, treatment and recovery
Waste (England and Wales) Regulations (2011)	Introduces a system for waste carrier registration

DID YOU KNOW

If waste is not managed properly and the duty of care is broken, then a fine of up to £5000 may be issued.

Safe methods of waste disposal

In order to dispose of waste materials legally, you must use the right method.

- Licensed waste disposal is carried out by operators of landfill sites or those that store other people's waste, treat it, carry out recycling or are involved in the final disposal of waste.

TRADE TIP

You can keep up to date with the latest developments in environmental law by using websites such as: www.environmentlaw.org.uk

REMEMBER

Metals, plastics, wood and cardboard can all be recycled.

- Waste carriers' licences are awarded to companies that transport waste; these may also be waste contractors or skip operators. For example, electricians or plumbers that carry construction and demolition waste would need to have this licence, as would anyone involved in construction or demolition.
- Recycling of materials such as wood, glass, soil, paper, board or scrap metal is dealt with at materials reclamation facilities. They sort the material, which is then sent to reprocessing plants so it can be reused.

Fig. 3.13 Recycled metals will be transported to a foundry

L2

- Specialist disposal is used for waste such as asbestos, which has to be double-bagged and placed in a covered, locked skip, along with any personal protective equipment (PPE) that may have become contaminated. There are authorised asbestos disposal sites that specialise in dealing with this kind of waste.

Approved processes for recycling materials

Metals

Scrap metal is divided into two different types:

- **Ferrous** scrap includes iron and steel, mainly from beams, cars and household appliances.
- **Non-ferrous** scrap is all other types of metals, including aluminium, lead, copper, zinc and nickel.

Recycling businesses will collect and store metals and then transport them to **foundries**. The operators will have a licence, permit or consent to store, handle, transport and treat the metal.

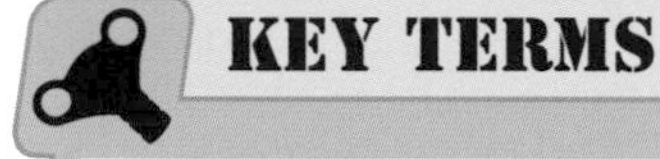

KEY TERMS

Ferrous: Metals that contain iron.

Non-ferrous: Metals that do not contain any iron.

Foundry: A place where metal is melted and poured into moulds.

L2

Plastics

Plastic waste can include drums, containers or plastic packaging. Different types of plastic are used for different things, so they will need to be recycled separately. Licensed collectors will pass on the plastics to recycling businesses who then remould the plastics.

TRADE TIP

You can search for licensed recycling and waste disposal companies using websites such as: www.wastedirectory.org.uk

Wood and cardboard

Building sites will often generate a wide variety of wood waste, such as off-cuts, shavings, chippings and sawdust.

Paper and cardboard waste can be passed on to an authorised waste carrier, but it is important to keep waste transfer notes for every load of waste passed on. The waste carrier must be registered or exempt from registration.

TRADE TIP

Before collection, plastics should be stored on hard, waterproof surfaces, undercover and away from water courses.

Disposal of potentially hazardous materials

Potentially hazardous waste materials need to be disposed of in the correct way. Legislation places a duty on individuals to ensure that hazardous wastes do not represent a risk to health or the environment.

Asbestos

Asbestos needs to be double wrapped in approved packaging, with a hazard sign and asbestos code information visible. The standard practice is to use a red inner bag with the asbestos warning and a clear outer bag with a carriage of dangerous goods (CDG) sign.

REMEMBER

Sites must only pass waste on to an authorised waste carrier, and it is important to keep records of all transfers.

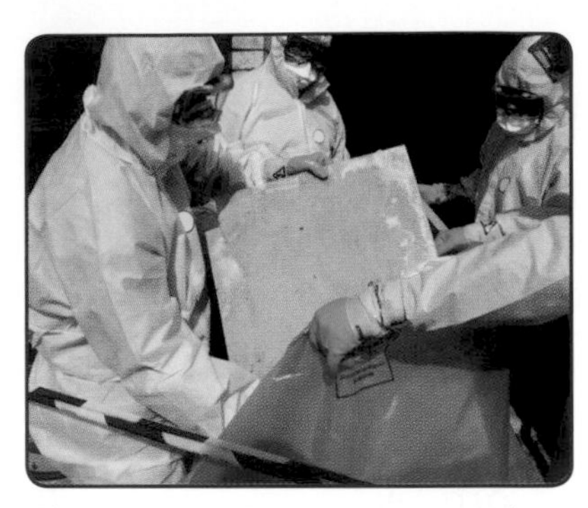

Fig. 3.14 PPE is essential when dealing with asbestos

REMEMBER

Asbestos waste includes anything that cannot be decontaminated, including PPE and cloths.

L2

Asbestos waste should be carried in a sealed skip or in a separate compartment to other waste. Ideally, it should be transported by a registered waste carrier and disposed of at a licensed site. Documentation relating to the disposal of asbestos must be kept for three years.

Electrical and electronic equipment

The WEEE Regulations were first introduced in the UK in 2006. They were based on EU law – the WEEE Directive of 2003.

Normally, the costs of electrical and electronic waste collection and disposal fall on either the contractor or the client. Disposal of items such as this are part of the Site Waste Management Plans Regulations (2008), which apply to all construction projects in England worth more than £300,000.

- For equipment purchased after August 2005, it is the responsibility of the producer to collect and treat the waste.
- For equipment purchased before August 2005 that is being replaced, it is the responsibility of the supplier of the equipment to collect and dispose of the waste.
- For equipment purchased before August 2005 that is not being replaced, it is the responsibility of either the contractor or client to dispose of the waste.

KEY TERMS

Fluorinated gas: A powerful greenhouse gas that contributes to global warming.

Refrigerants

Refrigerators, freezer cabinets, dehumidifiers and air conditioners contain **fluorinated gases**, known as chlorofluorocarbons (CFCs). CFCs have been linked

L2

with damage to the earth's **ozone layer**, so production of most CFCs ceased in 1995.

Refrigerants such as these have to be collected by a registered waste company, which will de-gas the equipment. During the de-gassing process, the coolant is removed so that it does not leak into the atmosphere.

KEY TERMS

Ozone layer: A thin layer of gas high in the earth's atmosphere.

DID YOU KNOW

The ozone layer provides protection against skin cancer by absorbing harmful UV radiation from the sun.

Preventing damage to the environment

Construction sites can be a major source of environmental pollution or danger. Those working on construction sites, therefore, need to follow environmental legislation and establish good pollution control procedures in order to prevent problems such as:

- noise
- smoke
- odour
- dust.

We all need to consider how our work activities could endanger the environment. Bonfires, for example, are not permitted on most construction sites.

Each site should have a nominated person who is responsible for carrying out a risk assessment to identify potential pollution incidents. The risk assessment needs to incorporate:

- how to stop the pollution occurring in the first place
- how to contain any pollution and prevent it from spreading
- information on who needs to be informed of pollution incidents

TRADE TIP

To deal with the problem of dust, use fine water spray or a sprinkler system that is capable of reaching all parts of the site.

TOOLBOX TALK

If there is a pollution incident on site you must contact the Environment Agency immediately.

- consideration of how significant a pollution incident might be
- plans for how to clean up and dispose of any pollution or waste.

Conserving water and reducing water wastage

Water is a precious resource, so it is vital not to waste it. To meet the current demand for water in the UK, it is essential to reduce the amount of water we use and to recycle water where possible.

The construction industry can contribute to water conservation by effective plumbing design and through the installation of water-efficient appliances and fittings. These include low- or dual-flush WCs, and taps and fittings with flow regulators and restrictors. In addition, rainwater harvesting and waste water recycling should be incorporated into design and construction wherever possible.

L2 Statutory legislation for water wastage and misuse

Water efficiency and conservation laws aim to help deal with the increasing demand for water. Just how this is approached will depend on the type of property:

- For new builds, the Code for Sustainable Homes and Part G of the Building Regulations set new water efficiency targets.

L2

- For existing buildings, Part G of the Building Regulations applies to all refurbishment projects where there is a major change of use.
- For owners of non-domestic buildings, tax reduction and grants are available for water efficiency projects.

In addition, the Water Supply (Water Fittings) Regulations (1999) set a series of efficiency improvements for fittings used in toilets, showers, washing machines, and so on.

Water efficiency calculator

To work out the water efficiency for a new build, you will need the water consumption figures provided by manufacturers for their products. This will determine the consumption of each terminal fitting. These figures are entered into a table to calculate the consumption of each fitting in litres per person per day. The total is then calculated, which needs to comply with the Code for Sustainable Homes and the Building Regulations 17.K.

The key points to remember are:

- Regulation 17.K requires the average water usage per person per day does not exceed 125l.
- Calculations need to be based on data for the fittings that have actually been installed.
- Completed calculations have to be sent to the building control body.

DID YOU KNOW

The water efficiency calculator for new dwellings can be found at the Planning Portal (www.planningportal.gov.uk), the government's online planning and building regulations resource for England and Wales.

Reducing water wastage

There are many different ways in which water wastage can be reduced, as shown in the following table.

Methods of reducing water wastage

Method	Explanation
Flow-reducing valves	Water pressure is often higher than necessary; by reducing the pressure, less waste water is generated when taps are left running
Spray taps	Fixing one of these inserts can reduce water consumption by as much as 70%
Low-flush WC	These reduce water use from 13 l per flush to 6 l for a full flush and 4 l for a reduced flush
Maintenance of terminal fittings and float valves	Dripping taps or badly adjusted float valves can cause enormous water wastage – a dripping tap can waste 5000 l a year
Promoting user awareness	Users who are on a meter will make savings if they improve their water efficiency, and their energy bills will reduce if they use less hot water

Captured and recycled water systems

KEY TERMS

Non-potable: Water that is unsuitable for drinking.

Greywater: Waste water from washing machines, sinks and baths or showers.

There are two variations of captured and recycled water systems:

- Rainwater harvesting captures and stores rainwater for **non-potable** use.
- **Greywater** reuse systems capture and store waste water from baths, washbasins, showers, sinks and washing machines.

Rainwater harvesting

In this system, water is harvested usually from the roof and then distributed to an above-ground or underground tank. Here it is filtered and then pumped into the dwelling for reuse. The recycled water is usually stored in a cistern at the top of the building.

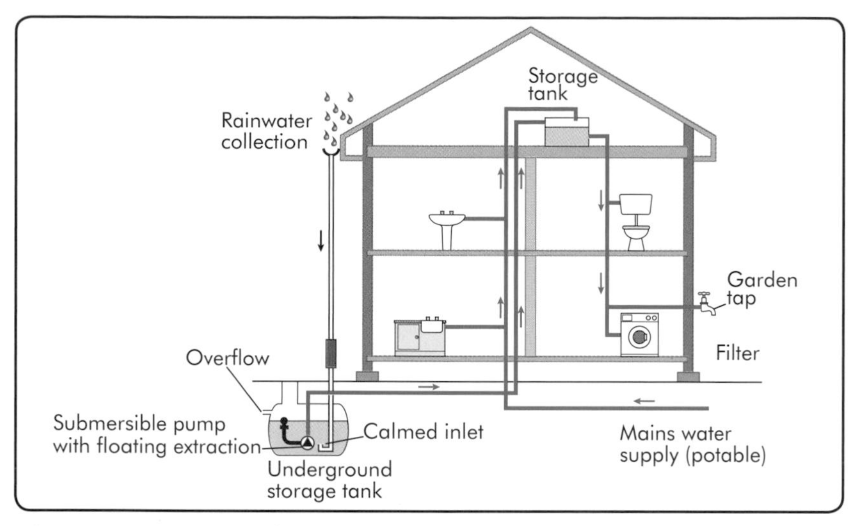

Fig. 3.15 A typical rainwater harvesting system

Greywater reuse

The idea of this system is to reduce mains water consumption. The greywater is piped from points of use, such as sinks and showers, through a filter and into a storage tank. The greywater is then pumped into a cistern where it can be used for toilet flushing, watering the garden or sometimes to feed washing machines.

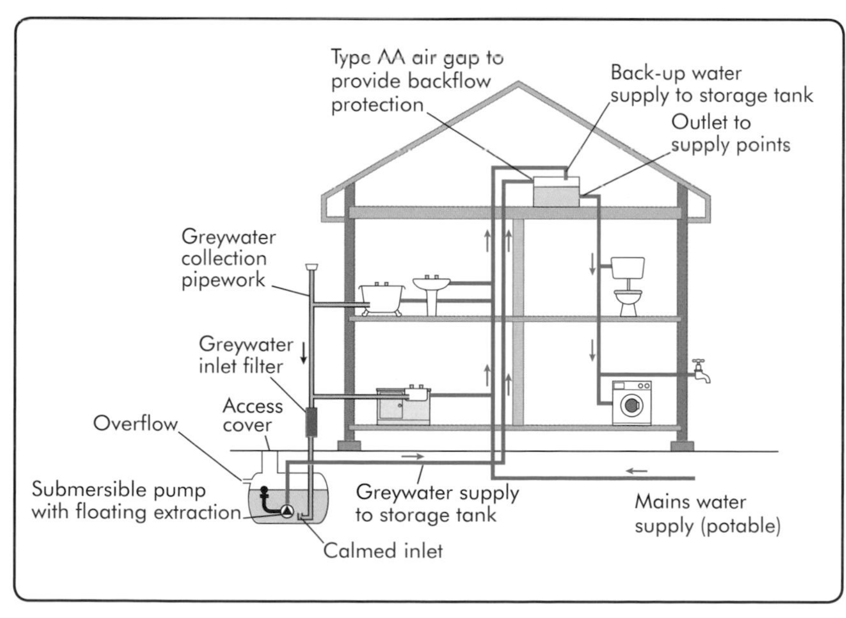

Fig. 3.16 Greywater system layout and key components

ACTIVITY

Fig. 3.16 shows the layout and key components of a greywater system. (It is not an installation diagram.)

Why might it be useful to have the greywater intlet filter above ground?

CHECK YOUR KNOWLEDGE

LEVEL 1

1. Identify and briefly describe two high carbon energy sources.
2. Identify and briefly describe two new technologies that are classed as low carbon energy sources.
3. What is the purpose of a DTC?
4. Briefly describe the basic principles of a solar photovoltaic installation.
5. Suggest and explain two ways in which you could reduce materials wastage.
6. List as many types of material you can think of that can be recycled from a normal building site.
7. How might a low-flush WC reduce water wastage?
8. Why might grewater be less useful as recycled water compared to rainwater?

CHECK YOUR KNOWLEDGE

LEVEL 2

1. Identify and explain one energy conservation law that is relevant to the building services industry.
2. Identify and briefly explain three zero carbon energy sources.
3. What is the main difference between CHP and CCHP?
4. Identify and explain three key parts of a system handover procedure.
5. Identify and explain two methods of conserving materials usage on site.
6. Which type of material needs specialist disposal?
7. Briefly explain the working principles behind a rainwater recycling installation.
8. Explain why mains water should not be contaminated by greywater or rainwater.

4 Scientific principles

This chapter covers the learning outcomes for:

Understand fundamental scientific principles within building services engineering

Understand how to apply scientific principles within mechanical services engineering

City & Guilds unit number 103 and L2 204; EAL unit code QACC1/03 and L2 QMES2/01; ABC A08 and L2 A08

Whichever sector in the building engineering services industry you work in, you will need to know some basic scientific principles. This is particularly the case if you are working within mechanical services engineering.

The basic scientific principles are all founded on clear and logical measurement, properties and reactions to different applications.

IN THIS CHAPTER YOU WILL LEARN ABOUT:

- standard units of measurement
- properties of materials
- energy, heat and power
- force and pressure
- simple mechanical principles
- principles of electricity.

Standard units of measurement

The International System of Units, or SI units, were developed in the 1960s. In the UK, SI units have not entirely replaced the old Imperial system, which includes pounds (lbs), feet (ft) and inches (in).

SI units commonly used in building services engineering

SI unit	Application and use
Metre (m)	A measure of distance. Metres are used in a wide variety of situations, such as the measurement of pipework, wiring, heights and widths of rooms and window apertures.
Kilogram (kg)	A measure of mass. 1 kg is almost equal to the mass of 1 l of water. Mass is often referred to as the weight of an object. It is therefore used in a huge number of situations, for example, measuring the weight of materials such as aggregates or a piece of equipment.
Second (s)	A measurement of time. A second is a building block of time – 60 seconds equals 1 minute, and 60 minutes equals 1 hour. Seconds can be valuable in calculating **velocity**, drying times or how long a part of a task takes to complete.
Kelvin (K)	A measure of temperature. 273.15 K is equal to 0 °C. However, 1 K is the same size as 1 °C, which makes conversion from one to the other very easy. For example, since 0 °C equals 273.15 K, 2 °C is equal to 275.15 K and 4 °C is equal to 277.15 K, etc.

KEY TERMS

Velocity: The speed of an object in a certain direction.

Molecule: Two or more atoms held together by strong chemical forces (bonds).

SI-derived units

The four main SI units of measurement form the basis of other ways of measuring, as can be seen in the following table.

The application and use of SI-derived units

SI-derived unit	Application and use
Square metres (m^2)	Area is measured in square metres. It is valuable to be able to work out the overall area of the floor of a room or the size of a plot of land. In other cases, the roof area would be worked out to calculate the amount of rainwater that could be harvested.
Cubic metres (m^3)	Volume is measured in cubic metres. You can work out the actual size of a room, taking into account its height, length and width. It is often used to calculate the amount of liquid in a container, such as a storage tank or cistern.
Litres (l)	Capacity is measured in litres. It is applied to liquids and can show, for example, how much liquid a cistern or drainage system can cope with or hold. It can also be used as an alternative measurement for the amount of materials that might be needed, such as paint or solvents.
Kilograms per cubic metre (kg/m^3)	The density of an object is measured in kilograms per cubic metre. Different materials have different densities. Density can show how different materials will interact if they are mixed together. It could be used to calculate, for example, the best materials to use for thermal insulation or for ballast.
Metres per second (m/s)	Velocity is measured in metres per second. It measures the rate and direction of movement. It is particularly useful for measuring the ability of pipework to cope with the flow of liquids or gases. By comparing velocity at different points, you can see whether a gas or liquid is speeding up or slowing down in a system.

ACTIVITY

You need to work out the volume of a room that is 6 m long, 4 m wide and 2 m high. You multiply the length by the width to get 24 m^2, and then multiply 24 by 2 to find the total volume, which is 48 m^3.

1. Using the same principles, what is the volume of a room that is 5 m long, 4 m wide and 3 m high?
2. Which building services engineers would be interested in this calculation, and why?

DID YOU KNOW

The Kelvin scale is an absolute thermodynamic temperature scale. Absolute zero is 0 K (−273.5 °C). It is a theoretical temperature where even molecules would stop moving.

Properties of materials

Materials used in building services are chosen for their specific properties. They may be chosen for their strength, hardness, ability to bend, how quickly they break down or whether they can conduct heat or electricity.

By knowing the properties of particular materials, you will be able to understand:

- why these materials are used for particular purposes
- what it is about them that makes them ideal for that use.

L2 Relative density

Relative density is often referred to as specific gravity. It means comparing one substance to another substance. It makes sense to compare the relative density of one gas with another gas, such as air, but you can also compare a gas to water, for example.

Relative density is almost always worked out by comparing the substance either with air or with water.

Water

To make the comparison, you begin by giving the relative density of water the value of 1.

- If the relative density of the other substance is less than 1, it will float on water. For example, an ice cube floats on water because its relative density is 0.91.
- A material such as steel has a relative density of 7.82, which means that it will sink in water.

L2 Air

The same procedure used for water is applied to gases. The relative density of air is also 1.

- The relative density of ammonia is 0.59, making it more buoyant than air.
- The relative density of carbon dioxide is 1.51, making it less buoyant than air.

Calculating relative density

In order to work out density, you divide the mass of the substance by its volume. To work out the relative density or specific gravity, you then divide that density by the density of either water or gas, whichever you are comparing it to.

For example, 1 m^3 of water at 4°C has a mass (weight) of 1000 kg. Remember that water has a relative density or specific gravity of 1. This helps us make a very easy calculation to work out the relative density of steel:

- 1 m^3 of steel has a mass (weight) of 7700 kg.
- Therefore the relative density of the steel is 7700 divided by 1000, or 7.7.
- This shows us that steel is 7.7 times denser than water.

ACTIVITY

1 m^3 of dry sand with gravel has a mass of 1650 kg. What is its relative density to water?

Solids

There are several different types of solid materials that are used routinely in the building services industry. They fall into three main categories:

- Metals
- Plastics
- Fire clays and ceramics

You need to be able to identify these and know how they are used.

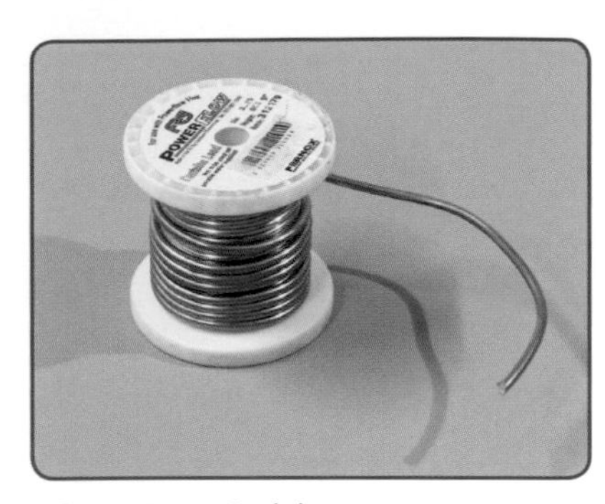

Fig. 4.1 Solder is an example of an alloy metal

Metals

Metals are broken down into three groups:

- Pure metals, such as aluminium or copper, only contain one type of metal.
- Ferrous metals always contain iron. Prime examples include the various types of steel, but steel is also an alloy (see below), as it is mostly iron with some carbon.
- Alloys are a mixture of two or more metals. A good example is solder, which is used to joint together metal work because it has a lower melting point than the metal that it is fixing together.

Some commonly used metals and their applications

Metal	Applications
Copper	It is used for pipework in plumbing and heating installations
Steel	An alloy of iron and carbon, it is also used for plumbing Stainless steel has chromium and nickel content and is used for sink units
Lead	In the past it was used for pipework, so can still be found in older properties It is also used for weathering on buildings
Cast iron	Contains a small amount of carbon It used to be used for pipework, but today, it is usually used for decorative work
Brass	A mix of copper and zinc It is used for pipe fittings, screws and bolts, and electrical contacts
Solder	This is either a mix of lead and tin or of tin and copper It is used for electrical connections and jointing material
Bronze	This is a mix of copper and tin It is used for corrosion-resistant pumps and decorative items
Gun metal	A mix of copper, tin and zinc, it is used for below-ground, corrosion-resistant fittings.

Plastics

Plastics, or polymers, are made from a substance found in crude oil called ethane. Ethane is itself made up of carbon, hydrogen and oxygen.

Ethane has the ability to make long polymer chains and makes polyethylene when it is heated under pressure. This is really useful for mouldings, and so plastics are used in a wide variety of applications, including pipework.

L2 There are two main types of plastic:

- Thermoplastics can be softened up if they are heated. They are poor heat conductors and they can be affected by sunlight, but they have a strong resistance to acid and alkalis. Examples include most pipework, including that used to channel boiling water.
- Thermo-setting plastics are used for moulding. They soften up when heated the first time and are moulded into a fixed shape, after which point additional heating will not change their shape. Examples include WC cisterns and plastic baths.

Plastics tend to be used for the majority of pipework. Specific types, like polyvinylchloride (PVC) are the most common types used for pipework. Unplasticised polyvinylchloride (UPVC) is commonly found in double glazing units.

DID YOU KNOW

PVC was originally created in 1835, but it wasn't until 1926 that it became more flexible and easier to produce.

Fireclays and ceramics

Fireclays and ceramics are materials or products that are made by baking (firing) various mixtures of sand, clay and other substances such as minerals. They can

Fig. 4.2 Chimney pots are an example of a mix of baked clay, in this case to create earthenware

produce a wide range of materials that are used in the building services industry. These include:

- roofing tiles
- earthenware chimney pots
- bricks
- traditional sinks and baths
- floor and wall tiling.

This class of solids also includes:

- mortar, which is made up of cement mixture, sand and water
- concrete, which is made from sand, gravel, water and cement.

ACTIVITY

1. What properties of bricks make them an ideal building component?
2. What are the advantages of using bricks over other types of building materials, such as wood?

Properties of solids

Solid materials are used in the building services engineering sector for their specific properties. The following table outlines the properties of solid materials, explains what they mean and gives examples of solids with these properties.

Properties of solids

Property	Meaning	Examples
Strength	Strength is the ability of a material to withstand a load or stress without breaking. Strength can be: ■ tensile, which means how much the material can be stretched ■ compressive, which is how much force or load it can bear.	■ A cable on a crane, or a tow rope, would need tensile strength, as it is under constant stretching force. ■ Flooring and roof joists need compressive strength, as they will take the weight of heavy objects and materials.
Hardness	Hardness is usually measured on a scale of 1 to 10. It is a measure of a material's ability to resist damage, such as being deformed, scratched or buckled.	Hardness is important for machinery parts because they need to resist wear and tear. Drill bits and circular saws are diamond tipped, as diamonds rate 10 on the hardness scale.
Ductility	Being ductile is the ability of the material to be bent out of shape without breaking. Some materials need to be very ductile and still retain their properties and integrity.	Copper is a ductile material that is used to make fine wires for electrical circuits. The material needs to be distorted into a very long, thin wire and still not break.
Malleability	This is the ability of a substance to be flexible enough to be worked without breaking.	Lead is a good example of a malleable material, as it is a soft metal. It can be cut, shaped and hammered to make flashing without any danger of the lead fracturing.
Conductivity	There are two aspects to conductivity – heat and electricity: ■ Heat, or thermal conductivity, is how poor or well a material allows heat to pass through it. ■ Electrical conductivity is how poorly or well the material allows electricity to pass through it.	■ Most metals are good conductors of heat. This is why you are likely to burn your fingers if you heat up one end of a copper pipe and then touch the other end. ■ Rubber-soled boots provide some protection from electricity, as rubber is a poor conductor of electricity.

L2 Why solid materials break down

Solids will break down or fail over time, depending on the environment or punishment that they take.

KEY TERMS

Oxidise: A chemical process that adds oxygen to a substance.

Ultraviolet radiation: Electromagnetic radiation, which we cannot see but that has a heating effect (e.g. as sunburn).

Molecular bond: A force that joins together the atoms within a molecule.

Fig. 4.3 Oxidisation creates iron oxide, or rust

TRADE TIP

Rust is atmospheric corrosion. It can be caused by pollutants, salt in the air, rain or humidity.

L2 Atmospheric corrosion

Metals can **oxidise** over time. Oxidisation occurs when the metal loses one or more electrons to create a metal oxide. For iron, this metal oxide is rust.

UV damage to plastics

Ultraviolet radiation from the sun degrades plastics by damaging the **molecular bonds** in the structure of the plastic. There are three types of ultraviolet radiation – UVA, UVB and UVC – but it is UVB that usually does the most damage to plastics.

Heat damage to plastics

Heat can break down the polymer chains in plastics, causing the material to fall apart or crumble. For this reason, plastic that is likely to be exposed to heat needs to be heat stabilised. This process involves adding an antioxidant during the manufacturing process.

Electrolytic corrosion

When a metal is exposed to water, it can dissolve or **ionise**. This affects the properties of the metal, and it no longer functions in the way that was originally intended. Corrosion can cause structural failure, leaks and loss of capacity. In effect, the metal is being destroyed by the electrochemical reaction. All corrosion is an irreversible reaction. When the metal is iron, the process is known as rusting.

Fig. 4.4 shows the electromotive series of metals. This lists metals in decreasing order of reactivity with hydrogen **ion** sources, including acids and water. In other words, those metals at the top of the list will react

L2

quickly when in contact with hydrogen ion sources, and those metals at the bottom of the list will react very slowly. The list is useful because it shows which metal or alloy will protect another that is lower in the series.

The rate of electrolytic corrosion will depend on whether the water is acidic or hot. Acidity will speed up the corrosion. Fig. 4.4 shows that those metals that are lower down the list will destroy those further up the list. The further apart the materials are on the list, the quicker the corrosion will take effect.

Blocks of a more reactive metal, such as magnesium or zinc, can be connected by conducting cables. These conducting cables can reverse the oxidisation process and the block of zinc or magnesium is sacrificed to keep the other metal from rusting. However, the magnesium or zinc needs to be replaced before it dissolves.

CORRODED END Anodic or less noble
Zinc
Aluminium
Cadmium
Steel
Lead
Tin
Nickel
Brass
Bronzes
Copper
Nickel–Copper Alloys
Stainless Steels (passive)
Silver
Gold
Platinum
PROTECTED END Cathodic or most noble

Fig. 4.4 The electromotive series of metals

KEY TERMS

Ionise: When atoms or molecules become charged by losing or gaining electrons.

Ion: An atom or molecule with a positive or negative charge.

DID YOU KNOW

You can buy UV-resistant plastic, but it is considerably more expensive than conventional plastic products.

DID YOU KNOW

Water molecules are made up of hydrogen (H^+) and oxygen (O^{2-}) ions. The chemical formula for water is H_2O.

DID YOU KNOW

The Institute of Corrosion (www.icorr.org) calculates the cost of corrosion each year to be 4% of Britain's total revenue.

ACTIVITY

1. What kind of specialist qualifications would you need to be a corrosion industry specialist?
2. What kind of sectors of the building services industry would you work in?

KEY TERMS

Friction: Resistance to movement that is caused when one surface rubs against another.

L2 Erosion corrosion

Erosion corrosion is when the surface of a material is degraded. In copper water pipes, it is often caused by the rapid flow of turbulent water. The speed of the water and the turbulence inside the pipe can result in some corrosion where the inner surface is torn off, starting the process of erosion corrosion. Over time, the metal is attacked by the corrosive action of the water and it is also eroded as the product of corrosion is removed from the metal surface.

Methods of preventing corrosion

Corrosion can be slowed down by:

- applying a coat, such as paint or enamel, or plating it with another metal
- applying a reactive coat, such as a corrosion inhibitor
- anodisation – a surface treatment that forms a hard surface layer
- applying a bacterial film (biofilm)
- using a sacrificial block – a more reactive metal that protects the main metal from corrosion or rusting.

Properties of liquids

Just as solid materials are chosen for their properties so, too, are liquids. The properties and applications of different liquids are described in the following table.

L2 ***Properties and applications of liquids***

Liquid	Properties and applications
Water	Water is a compound consisting of hydrogen and oxygen. It is a solvent, so gases and some solids can dissolve in it to form solutions. Water is vital as a natural resource in all buildings and dwellings. It also has practical applications, for example, it is used to create construction materials such as mortar or cement.
Refrigerant	Refrigerants can include ammonia, chlorofluorocarbons (CFCs), pure propane and hydrofluorocarbons (HFCs). Refrigerants' boiling points need to be below the target temperature. A refrigerant absorbs heat as it boils and vaporises. They are used in cooling systems.
Antifreeze/ glycol mixes	An antifreeze is a liquid that is added to water to lower the freezing point. Often ethylene glycol is used. It can be found in solar water heaters and chillers. The idea is that the antifreeze reduces the freezing point to below that of the lowest temperature that the system is likely to encounter.
Fuel oils	Fuel oils are produced via the distillation of petroleum. They are used to generate heat. One of the most common fuel oils is heating oil, or diesel, which is often used as a primary fuel source for hot water and central heating systems in areas that are not connected to mains gas. Fuel oils are complex liquid fuels, with hydrocarbons and small quantities of substances including nitrogen, sulfur and oxygen.
Lubricants/ greases	Lubricants and greases are ideal for reducing the **friction** between moving surfaces. Many lubricants are oil-based. They are used in most machinery, as they can help keep moving parts separate, thereby reducing friction and heat transfer and protecting against wear and corrosion.

Properties of water

Life on earth would be impossible without water, and it is the most common **compound** on earth. It is also an integral part of all construction work. Your work may involve routing water, disposing of it, protecting against it, collecting it or using some of its properties.

KEY TERMS

Compound: A substance made up of two or more elements that are chemically joined together, for example, water is hydrogen and oxygen.

Boiling and freezing point

Water's boiling point is 99.98°C, or 373.13 K, and its freezing point is 0°C, or 273.15 K.

TOOLBOX TALK

If water reaches a temperature of 100°C, it will become superheated. Few systems are capable of coping with this enormous pressure.

REMEMBER

Water is made up of hydrogen and oxygen (H_2O).

DID YOU KNOW

Water covers 70% of the earth's surface and the human body is almost 78% water.

KEY TERMS

Gravity: The force of attraction between objects, which is very small, so is usually only felt for very large objects such as the earth.

L2 Change of state and molecular changes

You will encounter water in three different states:

- At freezing point water is solid (ice)
- Above 273.15 K water is liquid
- At 373.13 K water becomes steam (water vapour), which is a gas.

As a liquid, water's maximum density is at a termperature of 4 °C, but when heated it will expand by up to 4%. This is because the heat energy makes the molecules move around and they distance themselves from one another so the water becomes less dense.

If water freezes it will expand by about 10%. This is why water that is stored in an enclosed space, such as a pipe, can cause the pipe to burst if the temperature drops to below freezing. Ice is less dense than water, which is why it floats on water (see page 96).

Boiling water will change into steam (water vapour). The water will expand extremely quickly, by as much as 1600%, as it becomes less dense. This is why stored water must be kept at temperatures below 100°C, otherwise the container, such as a hot water tank or cylinder, could explode under the pressure.

Capillarity

Water will rise in narrow tubes or pipes against the force of earth's **gravity**. This is because the water molecules are attracted to the solid material and to one another. This is known as cohesion. Cohesion creates surface tension, and the combination of the two allows the

L2

water to be drawn up the tube. The narrower the tube is, the greater the rise in the water level.

Acidity and alkalinity (pH)

Not all water is exactly the same. The nature of the water depends on where it came from or what it has been exposed to.

Water can be either acidic (soft water) or alkaline (hard water). The nature of the water is measured by its **pH** value. The pH scale goes from pH 1 (high acidity) to pH 14 (high alkalinity), with pH 7 being neutral (neither acidic nor alkaline).

KEY TERMS

pH: A measure of the acidity or alkalinity of a substance.

Rainwater is soft water as it is naturally slightly acidic. This is because, as it falls to earth, it passes through various gases and dust in the atmosphere. The slightly acidic rainwater will then fall onto the ground and may dissolve various substances, for example limestone.

DID YOU KNOW

A pH of less than 6.5 is acidic and corrosive. A pH of over 8.5 is hard water and will form scale and clog piping.

Hard water causes problems in direct hot water systems by depositing calcium carbonate, called scale, on the inside of pipes and boilers. There are two types of hardwater:

- Permanent hard water is alkaline and contains traces of the salt calcium sulfate.
- Temporary hard water is also alkaline and contains calcium carbonate, usually from chalk or limestone.

Applications of gases

Gases are widely used, particularly in the mechanical services industry. The following table outlines five of the most common gases and how they are used.

L2 *Five common gases and their applications*

Gas	Applications
Air and steam	Compressed air can be used for pneumatic tools, such as drills, nail guns, sandblasters and paint sprayers. Steam was originally used to run turbine engines, but today it is also used for cleaning buildings (as an alternative to corrosive acids).
Liquid petroleum gas (LPG)	LPG is widely used for gas torches, burners and heaters. Smaller cylinders are also used for blow torches. LPG is also used as a cooking or heating fuel in some dwellings that are not on the main gas grid.
Natural gas	Natural gas is primarily used for power generation. It is supplied to dwellings and other buildings via dedicated pipework, and provides energy for ovens, water heaters and central heating boilers.
Carbon dioxide	Carbon dioxide has direct uses in the industry as a compressed gas for pneumatic systems and pressure tools. It is found in some fire extinguishers and is sometimes used in welding. Liquid and solid carbon dioxides are also used as refrigerants.
Refrigerant gases	Refrigerant gases are used for air conditioning and include ammonia, sulfur dioxide, methane and carbon dioxide. Each refrigerant has its own characteristics and, therefore, uses, for example, R22 is found in most household refrigerators, while R134A is used in air conditioning and commercial refrigeration.

ACTIVITY

1. Find out what R22 and R134A are.
2. Which refrigerant is supposed to have the most impact on global warming?

L2 Properties of gases

Pressure and volume

Gases will always expand to occupy a container because their molecules are far apart from one another. So it is easy to compress a gas. If you double the pressure on a gas then the volume it takes up will reduce by about half, and you will, therefore, double its density. If you increase the temperature of a gas in a container, you will also increase its pressure.

All gases react in a fairly similar way, so there are scientific laws or equations that can help us understand the properties of gases.

L2 **Temperature of industry gases**

Gas	Boiling point (temperature they become gaseous, to the nearest degree at atmospheric pressure)
Water	100 °C (at atmospheric pressure) and at higher temperatures in higher pressures
LPG	**Ambient temperature** (the exact value varies with the propane/butane mixture)
Natural gas	−163 °C
Carbon dioxide	−79 °C
R22 chlorodifluoromethane	−41 °C
R134A tetrafluoroethane	−26 °C

Gas laws

One of the easiest ways to understand gas laws is to think about a sealed balloon. If you were to squeeze the balloon then two things would happen:

- air pressure in the balloon would increase
- the density of the air in the balloon would also increase.

Density is a combination of mass and volume. Since we know that the mass will stay the same in the balloon because the air cannot get out, by squeezing the balloon the density will rise and the volume will decrease, making the pressure go up.

KEY TERMS

Ambient temperature: The temperature in a room, or the temperature that surrounds an object.

Boyle's law

Boyle's law states that, at a constant temperature, the volume of a given mass of gas will vary inversely with pressure. In other words:

- the higher the pressure, the lower the volume
- the lower the pressure, the higher the volume.

Fig. 4.5 The gas laws work for hot air balloons

ACTIVITY

1. Imagine putting a balloon into a refrigerator. What do you think would happen and why?
2. Which law would this follow?

L2 Charles's law

Charles's law states that, at a constant pressure, the volume of a given mass of gas is directly proportional to its absolute temperature. In other words:

- density increases with a rise in temperature
- density decreases with a fall in temperature.

So, for example, if you warm up the gas molecules inside a balloon, they will speed up and start to move around more, hitting the skin of the balloon with increasing force. As the balloon's skin is elastic, the extra force will cause it to expand. In this way, the same mass of gas inside the balloon will take up a greater volume as the temperature increases.

Heat pump and refrigeration cycle

The heat pump and refrigeration cycle, shown in Figure 4.6, works in the following way:

1. Low-temperature heat enters the heat exchanger or evaporator. Heat is transferred from the source into the refrigerant, causing it to evaporate.
2. The refrigerant, which is now a gas, enters the compressor, and the pressure of the refrigerant is increased. Its temperature increases.
3. The refrigerant passes into the condenser, which is also a heat exchanger. It transfers the higher temperature heat into the air or water distribution circuit.

L2

4. The refrigerant, which is now cooler, enters the expansion valve. This reduces its pressure and temperature, returning it to its original state at the evaporator.

The cycle is then repeated.

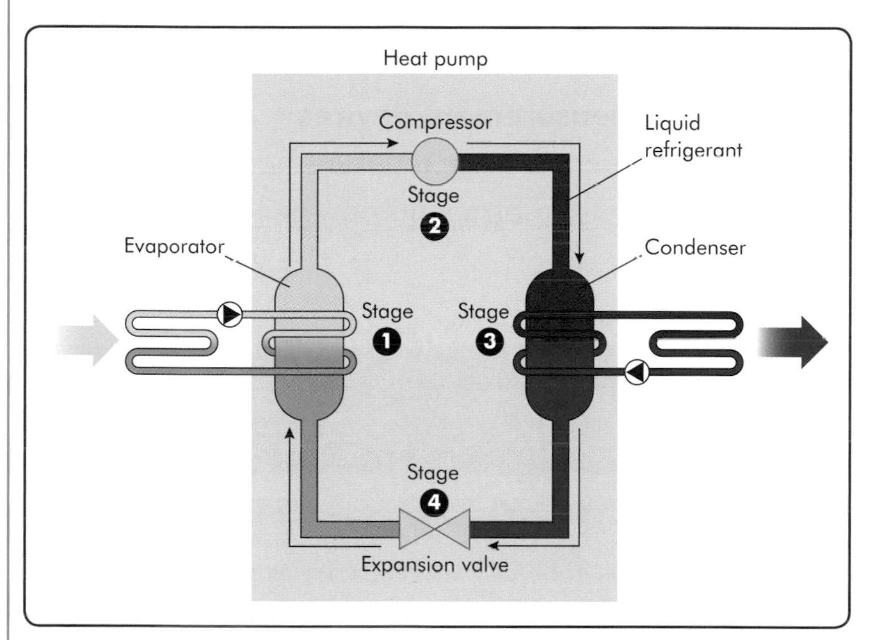

Fig. 4.6 Heat pump and refrigeration cycle

Energy, heat and power

Energy, heat and power are all concepts that you will encounter in building services engineering, so it is important to understand their relationship. In addition, you need to know how to carry out simple heat, energy and power calculations.

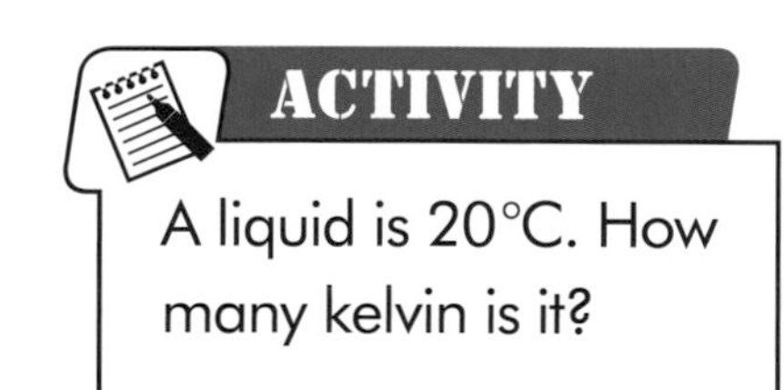

A liquid is 20°C. How many kelvin is it?

Celsius and kelvin

The most common measure of temperature is Celsius, but the SI unit is kelvin. 0°C is equal to 273.15K.

Since 1 K is equal to 1 °C, conversion from one to the other is very easy. Simply add 273.15 to the number of degrees Celsius to find out the equivalent kelvin measurement.

Temperature measurement devices

The majority of temperature measurement devices are thermometers, as can be seen from the table below.

Devices for measuring temperature

Temperature measurement device	Description
Bi-metallic strip or thermostat	These are found in ovens. They work on the principle that metals expand and contract at different rates. The metallic strip will bend when a particular temperature has been reached. This will then break the electric circuit and stop the oven from heating further.
Thermometer	These are based on the principle that mercury and alcohol expand or contract at a particular rate in response to changes in temperature. When these liquids are placed into a glass tube, the temperature can be read off a scale.
Pipe thermometer	These are often used for taking the surface temperature of pipes. They use bi-metallic strips and are particularly useful when a central heating system is being commissioned – one pipe thermometer is placed on the flow pipework and another is placed on the return pipework.
Digital thermometer	Essentially this is a probe, but it can also be connected to a **thermistor** to check the temperature of surfaces.

Changes of state

By adding energy to or taking energy away from a substance, it is possible to change its state to solid, liquid or gas. This is because every substance is made up of particles.

- Heating a solid makes the particles move more and the solid can become a liquid (it melts).
- Heating a liquid makes the particles move faster, causing the liquid to evaporate (turn into a gas).
- If you cool down a gas then its particles slow down and condense (turn into a liquid).
- If you cool down a liquid, the particles slow down further and eventually will freeze (become solid).

Thermistor: A device in which electrical resistance changes with temperature.

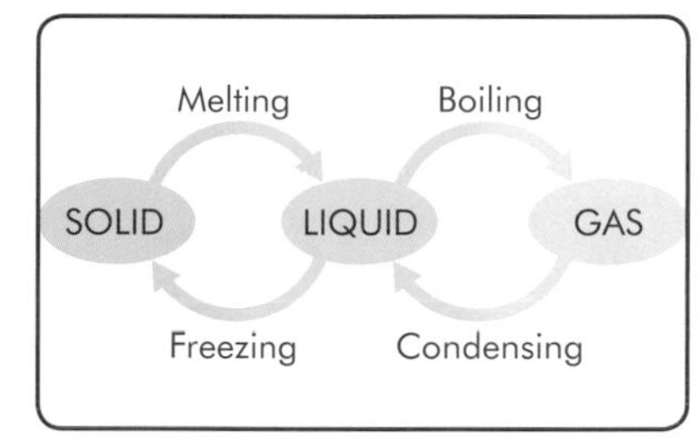

Fig. 4.7 Changes of state

REMEMBER

The temperature at which a liquid becomes a gas is its boiling point. The temperature at which a solid becomes a liquid is its melting point.

Heat transfer

The three ways in which heat can be transferred are conduction, convection and radiation.

Conduction

Conduction occurs in solids when heat is transferred through the material as a result of the molecules vibrating more. The vibrations are passed on through the material.

- Metals are good conductors of heat, for example, iron, lead and copper are highly conductive.
- Materials such as wood, plastic and ceramics are poor conductors of heat. We say that they are good **thermal insulators**.

Thermal insulator: Any material that is a poor conductor of heat.

REMEMBER

Conduction does not just apply to solids. Gases and liquids also conduct heat, though less efficiently.

KEY TERMS

Fluid substance: A gas or liquid, which can flow.

Convection

Convection occurs in gases and liquids, such as air or water, which are both known as **fluid substances**. When the fluid substance is heated, it expands, which means that it will have a lower density. The warm fluid rises and is then replaced by a colder and denser fluid below. The currents allow a continuous flow of upward heat, away from the source. This makes air and water ideal for convector heaters and for domestic hot water systems that use immersion heaters.

Radiation

Radiation is the transfer of heat from one body to a cooler one. The heat radiation is, in effect, heat waves.

Heat radiation is absorbed at different rates by different materials. For example, a polished surface will not absorb radiated heat as easily as a dull surface.

L2 Latent and sensible heat

Understanding latent and sensible heat is important.

- Latent heat is absorbed or given off by a substance as it changes its physical state. The heat that is absorbed does not actually cause a temperature change in the substance. For example, when water boils, it remains at 100 °C. Therefore, any heat added to keep the water boiling is latent heat, because it does not cause a temperature change.
- Sensible heat is absorbed or given off by a substance that is not in the process of changing its physical state. It can be measured using a thermometer.

L2

An important application of latent and sensible heat is the use of refrigerants in cooling systems. When the refrigerants absorb latent heat to evaporate in the system they cool the surrounding air.

Units of energy and heat

Energy is the ability to do work on a substance, for example, by pushing, pulling or lifting.

Heat affects substances at the molecular level and, as we have seen, can be transferred in three different ways – by convection, conduction or radiation.

There is an important distinction between temperature and heat:

- Temperature is the measure of the degree of hotness or coldness.
- Heat is the total energy associated with the motion of molecules.

DID YOU KNOW

Water in a boiled kettle has a higher temperature than warm bathwater. However, bathwater has more heat because its mass is greater.

Energy and joules

Energy is expressed in terms of joules (J). One joule (1 J) is equal to one watt (1 W) of power for one second (1 s).

Specific heat capacity

This is the amount of heat that is needed to raise the temperature of 1 kg of a particular substance by 1 °C. It is expressed as kJ/kg°C.

The specific heat capacity of materials differ. For example, the specific heat capacity of cast iron, a hard solid, is 0.554, whereas the specific heat capacity of water, as a liquid, is 4.186.

DID YOU KNOW

The unit of power (the watt) is named after James Watt, a Scottish engineer who also coined the term 'horsepower'.

Power

Power is shown as watts (W). It is equivalent to 1 joule per second.

L2 Heat, energy and power calculations

Heat, energy and power calculations are best understood by means of an example:

- You have 300 l of water, and you want to raise the temperature of the water from 10°C to 60°C.
- You know that 1 l of water weighs approximately 1 kg.
- The specific heat capacity of water is 4.186 kJ/kg°C.
- Heat energy = 300 l × 4.186 kJ/kg°C x (60°C − 10°C) = 62,790 kJ

In other words, you multiply the amount of water by its specific heat capacity and then by the difference between the two temperatures.

To work out the amount of power that you need in order to heat 300 l of water in 1 hour, you need to do a second calculation. (You have decided to assume that no energy is lost in heating the water.)

The first thing to note is that you are using kilowatt hours (kW/h). There are 3600 seconds in each hour (60 seconds times 60 minutes), so the calculation is:

$$\frac{62{,}790}{3600} = 17.44 \text{ kW}$$

ACTIVITY

You have 400 l of water that you want to raise in temperature from 20°C to 80°C. Work out the amount of heat energy and power required to heat this water in 1 hour.

Force and pressure

You need **force** to change an object's velocity or acceleration. **Pressure** is similar to force – it is the application of force over a particular area. An understanding of both of these concepts is important in the building services industry.

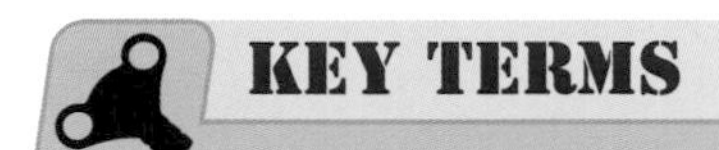

Force: The push or pull that acts between two objects.

Pressure: Force per unit area.

L2 Units of force and pressure

There are several key units of force and pressure that are all derived from SI units.

- Acceleration (*a*) is a measure of change in velocity over time (m/s^2). In SI units, this is the change in metres per second in 1 second.
- Force due to gravity, commonly known as weight, is the amount of force that pulls an object towards the earth, measured in newtons (N). The mass (*m*) of the object, in kilograms (kg), is multiplied by the acceleration due to gravity (*g*), which is approximated as $9.81\,m/s^2$. So the force due to gravity of a 1 kg mass is 9.81 N.
- Force is mass (*m*) × acceleration (*a*), and is measured in newtons (N). 1 N is the amount of force required to accelerate a mass of 1 kg at a rate of $1\,m/s^2$.
- Pressure is force per unit area and is measured in pascals (Pa). 1 Pa is equivalent to 1 N per square metre (N/m^2).
- Atmospheric pressure is the pressure exerted on the earth's surface by the weight of air in the atmosphere. It is dependent on altitude (distance above sea level), temperature and humidity. A siphon

Fig. 4.8 The motorbike may be moving fast, but if it is at a constant speed it is not accelerating

DID YOU KNOW

In building services engineering, you will also come across the term 'bar' when referring to pressure. 1 bar is 100,000 Pa, and is roughly equal to the atmospheric pressure on earth at sea level.

L2

uses atmospheric pressure. By forcing a quantity of fluid into a pipe and over the bend or crown, the water will continue to flow along the pipe from a higher container to a lower one owing to the difference in the weight of water, due to atmospheric pressure.

- Flow rate is measured in cubic metres per second (m^3/s). It is used for water flow and, in heating, ventilation and air conditioning, for air flow. 1 m^3/s is equivalent to 1000 l/s (also kg/s for water).

Application and use of units of measurement of pressure and flow rate

There are various ways in which pressure and flow rate can be measured. These usually depend on the application, as can be seen from the table below.

L2 ***Application and use of units of measurement of pressure and flow rate***

Unit of measurement	Application and use
Bar/millibar	1 bar is equal to 100,000 Pa. It is often used to specify the pressure in compressed gas cylinders.
Kilopascal (kPa)	1 Pa is equal to the force of 1 N/m^2. Since the pascal is a small unit, kilopascal is widely used. 1 kPa is equal to 1000 Pa.
Pounds per square inch (psi)	Largely replaced in the UK by the bar, this is still common in the USA.
Metre head (m of head)	This relates to the pressure of water. Water pressure measures the force needed to move water from the mains supply into the dwelling's pipes. 1 m of head is equivalent to a 1 m high column of water.

L2

Unit of measurement	Application and use
Cubic metres per second (m^3/s)	These are often used for water flow, especially in rivers, but are also used for measuring air flow.
Litres per second (l/s)	1000 l is equivalent to 1 m^3/s. 1 l of water has a mass of almost exactly 1 kg.
Kilograms per second (kg/s)	This is a measurement of flow rate derived from SI units. Because 1 l of water weighs 1 kg, kg/s and l/s are practically the same value for water flow.

Force and pressure calculations

Force exerts an influence on an object or substance, causing it to undergo a change in speed, direction or shape. Force has a strength and direction so is mass multiplied by acceleration.

Pressure is the force that is applied to the surface of an object. Pressure = force/area.

Simple force and pressure calculations

Calculation	How to do it
Pressure head	This is used to show the energy of a fluid due to the pressure exerted on its container. It is sometimes called static pressure head or static head. It is equal to the fluid's pressure divided by the specific weight, or the fluid's pressure divided by the density of the fluid and its acceleration due to gravity.
Static pressure	To work out static pressure, you need to multiply density by gravity by height. Water density is 1000 kg/m^3, gravity is 9.81 m/s^2 and height is the difference in levels.
Dynamic pressure (Q)	To work out dynamic pressure, you need to know the density of the fluid (*p*) and the velocity of the fluid (*v*). The formula is: $Q = \frac{1}{2} pv^2$

Velocity, pressure and flow rate

If you increase pressure, then the velocity and flow rate will fall. If pressure is reduced, then velocity and flow rate will increase.

If the pipe size is reduced and the flow rate remains constant, the velocity will increase.

Why pipework restricts flow

Pipework can restrict the flow of liquids and gases for a number of reasons:

- Change in direction, bends and tees – bends can cause the flow to separate from the wall of the pipe, wasting energy and reducing flow.
- Pipe size – if the pipe is too small then there will be friction inside the pipe, which will reduce the flow.
- Pipe reductions – without streamlining a reduction in pipe size, bubbles can form at the connection, which will reduce flow.
- Roughness of material surface – the roughness of the interior surface of the pipe can cause additional friction, leading to a loss in flow.
- Constrictions such as valves – control valves regulate the flow. As the fluid passes through a valve, there is a drop in pressure. The higher the flow rate through a valve, the greater the pressure drop.

Reductions in the size of the pipework or constrictions such as valves can be designed to reduce the flow rate. In other cases, there are unavoidable reasons why the flow rate is affected, largely related to the need to route the liquid or gas in a particular direction. The requirement to link a main pipe to subsidiary pipes, or the choice and suitability of the material that the pipe is made from, can also affect flow.

Simple mechanical principles

Perhaps the most important breakthrough in mechanical engineering occurred around 300 years ago, when Sir Isaac Newton created his laws of motion. The principles of basic mechanics are still largely based on Newton's work.

The principles behind simple machines

A simple machine is a mechanical device that changes the direction or magnitude of a force. It uses mechanical advantage or leverage to multiply force.

Simple machines are the building blocks of much more complicated machinery. The table on page 122 outlines the key principles behind some of them.

REMEMBER

Friction is resistance to movement that is caused when one surface rubs against another.

The mechanical principles of simple machines

Simple machine	Mechanical principle
Mechanical advantage (MA)	MA is a measure of the force gained by using a tool, mechanical device or machine system. For example, if you try to undo a nut with your fingers, a huge amount of force is needed. But if you use a spanner to undo a nut, less force is needed. This is because the spanner increases the distance between the fulcrum and the line of action of the force, which multiplies the turning effect of the force on the nut. When calculating MA, the effort arm is where force is applied. The effort arm is always larger than the resistance arm. MA = length of effort arm ÷ length of resistance arm
L2 Velocity ratio	This is the ratio of the distance moved by the effort applied to the load, to the distance moved by the load itself. In the case of an ideal (frictionless and weightless) machine, velocity ratio equals MA. Velocity ratio is also called distance ratio.
Levers	There are three classes of lever: ■ The simplest lever is a rigid bar that can be turned freely round a fixed point (fulcrum), such as a see-saw. The first class of levers has the fulcrum (pivot) in between the resistance at one end and the effort at the other. If the resistance is great, then the fulcrum must be nearer to it than to the effort. ■ The second class of levers has the fulcrum at one end and the effort at the other with the resistance in the middle, such as a wheelbarrow. ■ The third class of levers needs a greater effort than the amount of resistance moved. The fulcrum is at the end and the resistance at the other end with the effort in the middle. For instance, the human arm uses this method with the elbow in the middle.
Wheel and axle	This is a first-class lever – basically a rod attached to a wheel. The wheel and axle can be used as a force multiplier (e.g. a door knob) or as a distance multiplier (e.g. a bicycle). When the axle is turned, the outside of the wheels turn at a greater speed proportional to the ratio of the radius of both the wheel and the axle.
Pulleys	This is a wheel on an axle or shaft. It can be used together with ropes, cables, belts or chains. These run over the wheel inside a groove. The pulley changes the direction of an applied force. The MA is calculated by the number of rope lengths exerting force on a load.
Screws	These convert rotational movement into linear movement. In other words, rotating a screw forces it into a block of wood. Each full turn of the screw, or lead, creates a MA. The smaller the lead, the higher the MA.

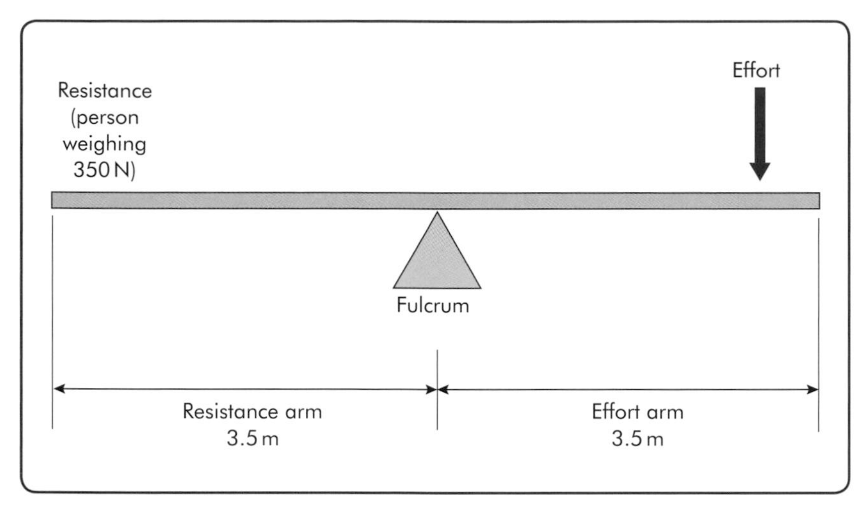

Fig. 4.9 A lever

a) First-class lever
Output
Input

b) Second-class lever
Output
Input

c) Third-class lever
Output
Input

Fig. 4.10 The three classes of lever: a) first order; b) second order; c) third order

ACTIVITY

Look at the lever shown in Fig. 4.9.

To find the MA of a lever, you need to divide the effort arm length by the resistance arm length.

$$MA = \frac{\text{effort arm length}}{\text{resistance arm length}}$$

What is the length of the resistance arm and the effort arm in Fig. 4.9?

Principles of basic mechanics

Mechanics is all about the behaviour of objects when they are subjected to force. Many of the ideas, particularly those of Newton, are classical mechanics, as the following table shows.

The principles of basic mechanics

Principle	Explanation
Theory of moments	A **moment** is a turning force. It is a force (*F*), acting at a perpendicular distance (*d*) from the turning point, so the moment of a force is $F \times d$. Moments are measured in Newton metres (Nm). They can act in two ways – clockwise or anticlockwise. When more than one force acts in the same direction, the overall turning force is the sum of their moments: $(F_1 \times d_1) + (F_2 \times d_2)$ When forces act in opposing directions, in order for them to balance, the total turning effect in each direction must be the same (clockwise moment $F \times d$ = anticlockwise moment $F \times d$).
Action and reaction	Force always operates in pairs. For every force acting on an object, there is an equal and opposite reaction force. We call these forces the action–reaction pair. For example, your weight pushes down on the floor and the floor pushes up against you with an equal force.
Centre of gravity	This is applicable to volume, area or line, and is the point at which the object would be in balance if suspended.
Equilibrium	The forces acting on an object are said to be in equilibrium when they are balanced. In other words, there is no resultant force acting on the object.

Moment: The turning effect of a force.

Principles of electricity

We have become incredibly reliant on electricity, although it is something that we cannot see or hear. You can be seriously injured or killed if electricity travels through you, so you need to always handle it with care.

Electron flow theory

Electricity is the flow of charged particles, which can be either electrons or ions. As far as physics is concerned, the focus is on electricity as a flow of electrons.

In a circuit, an electrical charge will flow from the cell or battery along a wire to a lamp or light bulb and then back to the cell. Cells are usually drawn with a long line and a short line (see Fig. 4.11). The long line is the positive side and the short line is the negative side.

Circuit diagrams are drawn as though the electrical charge actually flows from positive to negative. However, electrons are negatively charged. This means that the true flow of electrons is from negative to positive. They are repelled from the negative side of the cell and attracted to the positive side.

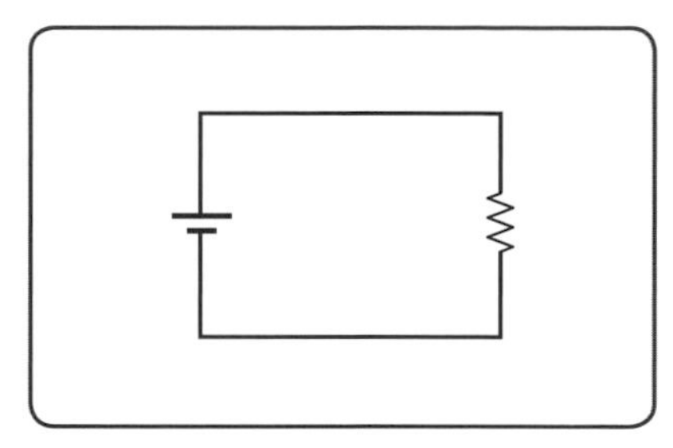

Fig. 4.11 Circuit diagram showing a cell and a light bulb

Measuring electrical flow

In order to work out the flow of electricity through a circuit, you need to know about voltage, current and resistance.

- Voltage, or potential difference (pd), is a measure of the energy available to drive the flow of electrons. It describes, in effect, the pressure that pushes electrons along a circuit. Voltage is measured in volts (V).
- Electrical current (*I*) is the flow of electrons between two points. It is measured in amperes (A), which can be shortened to amps.
- Resistance (*R*) is anything that can slow down the flow of electrons. It is measured in ohms (Ω).

DID YOU KNOW

A circuit that flows from positive to negative shows conventional current. Real current is the actual flow of electrons from negative to positive.

Material conductivity and resistance

Some materials are good **conductors** of electricity – that is, they allow an electrical current to pass through them easily. Copper, aluminium and silver are all excellent electrical conductors, so they are used for the wiring in electrical circuits. (The wiring also needs to be the right diameter.)

KEY TERMS

Conductor: Any material through which an electric current will flow.

KEY TERMS

Resistance: The way in which a material prevents the flow of electric current.

REMEMBER

In order to encourage the flow of electrons in a circuit, the materials used must be electrically conductive.

DID YOU KNOW

The UK's electricity supply frequency is 50 Hz (hertz). This means that there are 50 AC cycles each second.

KEY TERMS

Joule: The equivalent of passing 1 A of electric current through 1 Ω of resistance for 1 s.

Other materials are poor conductors of electricity and slow down the flow of electrons through a circuit. We describe these materials as having **resistance**.

Devices with high resistance are often placed in a circuit in order to reduce or control the flow of current. For example, resistors can be used to control the flow of electricity to a kettle, to protect it from receiving an overload.

Direct and alternating current

Alternating current (AC) is the standard type of current. It is used in domestic dwellings and the majority of other buildings in the UK. The term alternating refers to the flow of the electrons. The electrons flow in cycles, or waves. For the first half of the cycle the electrons flow in one direction and in the second half they flow in the opposite direction.

Direct current (DC) is when the electrons in a circuit flow in the same direction at all times. Batteries, torches and power tools use direct current. DC generated by batteries in a boiler control circuit, for example, need a transformer in order to reduce the voltage and convert it to AC.

Units of electrical measurement

You need to be familiar with four simple units of electrical measurement, as shown in the following table.

The purpose and application of units of electrical measurement

Unit of electrical measurement	Purpose and application
Current (amperes)	This is a measurement of the flow of electrically charged particles, or electrons. It allows us to measure how much electricity is flowing in a given circuit at any particular time.
Voltage (volts)	This is a measurement of electromotive force (emf). It is the number of **joules** required to push one **coulomb** of electrons around a circuit.
Resistance (ohms)	This is the degree to which either a device or material resists the flow of electrical current.
Power (watts)	This measures the rate of energy conversion. 1 watt is equivalent to 1 joule per second.

Simple electrical equations

You have already seen that there is a relationship between voltage, current and resistance. This forms the basis of the first simple electrical calculations that you need to understand. The following table outlines all four of these calculations.

Coulomb: The unit of electrical charge. 1 C is equivalent to a flow of 1 A in 1 s.

Four simple electrical calculations

Calculation	Explanation
Ohm's law	Ohm's law allows us to work out voltage (V), current (I) or resistance (R) if these are unknown, provided we know the value of the other two: ■ To work out voltage, multiply the current by the resistance ($V = IR$) ■ To work out current, divide voltage by resistance ($I = V/R$) ■ To work out resistance, divide voltage by current ($R = V/I$)
Power consumption of electrical circuits	A 100W light bulb that has been left on for 10 hours uses 1 unit of electricity – 1 kWh (kilowatt hour). This is calculated by multiplying the power requirement of the device by the number of hours used. In this case: 100 × 10 = 1000 Wh (watt hours), or 1 kWh.

Calculation	Explanation
L2 Over-current detection device size	A fuse is an over-current detection device. To find the fuse rating for a particular appliance, you need to divide the power of the appliance in watts by the voltage of the electricity supply. So, for our 100 W light bulb, the calculation would be 100/230 V (this is the voltage of mains electricity) = 0.434 A. Fuses are of standard values, so a 1 A or 3 A fuse is usually sufficient to provide protection.
Voltage, current and resistance in series and parallel circuits	In a **series circuit**, the total resistance is worked out by adding together all of the resistances within that circuit. The supply voltage of the circuit is equal to each of the individual voltages across each resistor. In a **parallel circuit** (most power and lighting circuits in domestic dwellings), the total current is worked out by adding together all of the current flowing through each branch. The voltage will be the same in each branch and the total resistance can be discovered by using the formula: $1/R = 1/R_1 + 1/R_2 + 1/R_3$

KEY TERMS

Series circuit: Circuit in which the components share the current.

Parallel circuit: Circuit in which components share the energy source but not the current.

Circuit breaker: A safety device that interrupts an electric current.

L2 Earthing of electrical circuits

To prevent damage caused by potential overheating, all pipes, radiators, sinks, baths and electrical appliances need to be earthed. This means that a wire has to permanently connect them to a metal earthing block in the consumer unit. Then, should an electrical fault occur, the current will be carried away to the earthing block and the change in the electrical flow will blow either the **circuit breaker** or the fuse.

The earth wire looks slightly different in some cables:

- In cabling and circuits, the earth wire is an unsheathed copper wire.
- In appliances and for sinks, radiators, basins and pipework, the earth wire is yellow and green.
- For mains water and gas pipes, the earth wire should be within 600 mm of the meter or stopcock.

CHECK YOUR KNOWLEDGE

LEVEL 1

1. What is the SI unit for mass?
2. What two SI units are used for velocity?
3. How are fireclays and ceramics used in the building industry?
4. If something is said to be ductile, what does this mean?
5. What is the SI unit used for power?
6. How might a rough pipe restrict water flow?
7. What is meant by mechanical advantage?
8. Briefly explain Ohm's law.

CHECK YOUR KNOWLEDGE

LEVEL 2

1. Briefly explain the reasons for atmospheric corrosion in metals.
2. If something is said to be thermally conductive, what does this mean?
3. What is meant by temporary hard water and permanent hard water?
4. Briefly explain Boyle's law.
5. What units of measure are used for flow rate?
6. Briefly describe the mechanical advantage of a wheel and axle.
7. Briefly explain the difference between direct current and alternating current.
8. Why is it necessary to earth an electrical circuit?

5 Roles, responsibilities and procedures L2

This chapter covers the learning outcomes for:

L2 Understand the roles, responsibilities and procedures within building services engineering

City & Guilds unit number L2 203; EAL unit code L2 QACC2/09; ABC L2 A01.

The building services engineering industry is complex. There are many different roles, responsibilities and career opportunities within the industry. There is also a wide range of documents that are used in the workplace or provided to customers and clients. The sector also includes several different types of business or organisation, and this can determine the type of communication methods and document procedures that are used. There are also major differences between domestic and industrial or commercial contracts.

IN THIS CHAPTER YOU WILL LEARN ABOUT:

- building services engineering systems
- roles, responsibilities and career opportunities
- documents and documentary procedures
- types of businesses and companies.

Building services engineering systems

The term building services engineering covers the internal make-up of any building and its environmental impact. It includes:

- the design
- installations
- the operation and monitoring of any mechanical or electrical systems
- safety aspects
- environmental controls
- the long-term **sustainability** of the building.

KEY TERMS

Sustainability: In terms of building services engineering, this is about reducing a building's environmental impact over its lifetime.

Building environmental management

Environmental management systems are now a major part of construction, and are put in place from the very beginning. This is to avoid or minimise any harmful effects to the environment or surrounding communities.

As a project moves forwards from the initial design, actual construction issues, such as dates for particular types of work, working hours and a host of other considerations, need to be addressed. Depending on the type of construction, there may be other considerations:

- Heritage – is the building or surrounding area in some way protected as a historic site?
- Ecology – what will be the effects (direct and indirect) of the construction on the local plants and wildlife?

- Landscape – will the building affect the overall look of the area?
- Lighting – will the building cause light pollution?
- Noise and vibration – will work during the construction and during the normal operations of the building affect neighbouring buildings?
- Pollution – what safeguards have been put in place to prevent pollution during construction and afterwards?
- Traffic – what will be the impact during construction and in the longer term of additional vehicles in the area?
- Waste – how will construction waste and future waste generated by the users of the building be managed?

Building services engineering systems in different settings

There will be a range and mix of professionals involved in work on different types of sites or projects. All sites will need some kind of design or plan, and this is usually the responsibility of professionals such as architects, structural engineers or quantity surveyors.

The actual engineering work will depend on the scope or size of the project:

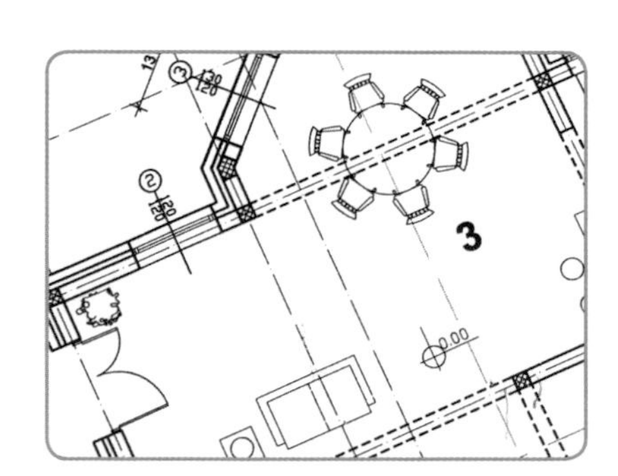

Fig. 5.1 Alterations to a public library would be carried out from a plan

- In a domestic building the work could be carried out either by an individual or a small team, possibly working to designs.
- Buildings used by businesses or the public, such as office blocks and hospitals, may have to be closed or at least partially closed while work is being carried out. In these instances, a main contractor would work to a

set of designs and would use a range of individuals or sub-contractors to carry out specialist work.

- For larger commercial and industrial buildings, such as factories or food processing centres, specialist equipment may need to be installed and some structural work carried out. Usually, larger contractors work on these projects and may use smaller sub-contractors or individuals when and where required.

It is likely that you will encounter a wide range of different systems used by the various professions within the building services engineering industry. Increasingly, you will come across systems that are designed specifically for energy-saving uses or to reduce the building's overall impact on the environment.

Fig. 5.2 The construction of large commercial or industrial buildings requires specialist equipment

Building environmental management

This is a relatively new, but growing area of building services engineering. The idea is to create new builds, or to **retrofit** existing buildings with materials and systems that are sustainable and have minimal impact on the environment.

Retrofit: Adding new technology or features to something that already exists.

Building environmental management aims to:

- reduce the overall impact of building construction on the environment
- reduce the long-term impact that a building has on the environment
- use renewable energy sources, such as solar or wind power, wherever possible
- use natural or low-impact products during construction

- conserve or reduce the use of water
- cut the amount of carbon produced during construction and the lifetime of the building
- reduce the need to use non-renewable energy sources
- cut down on waste during construction and the life of the building
- use recycled materials wherever possible.

There is an enormous scope for building environmental management systems. Some practical applications include:

- installing greener, renewable technology to heat dwellings and provide hot water
- recycling rainwater or reusing **greywater**
- installing natural insulation systems, such as wool, hemp, wood fibre, recycled plastic bottles or hay
- **eco-friendly** production of windows and doors, paints and wood finishes
- **smart technology** to reduce energy use in unoccupied parts of the building
- energy-saving monitors, which can show the energy use around the property (when used with smart plugs they can control individual appliances).

KEY TERMS

Greywater: Waste water from washing machines, sinks and baths or showers.

Eco-friendly: Having a minimal negative environmental impact.

Smart technology: Sensors that can detect whether a room or part of the building is occupied and then shut down systems, such as lighting, if the area is unoccupied.

There are a number of different environmental technology systems, which generally fall into four categories:

- Heat-producing – solar thermal hot water, heat pumps, biomass
- Electricity-producing – solar photovoltaic, micro-wind, micro-hydro

- Co-generation – microcombined heat and power
- Water conservation – rainwater harvesting, greywater reuse.

ACTIVITY

Find out about two of these environmental technologies and what a basic system would look like in a domestic dwelling.

Roles, responsibilities and career opportunities

The building services engineering industry is an incredibly wide sector, with a huge variety of professions, each with their own distinct career paths. Essentially, career and progression can be split into two main paths:

- Trade-based careers incorporate plumbing, electrical, heating and ventilation, and refrigeration and air conditioning. Usually, these are competence-based and many of the key skills are learned as an apprentice while studying vocational courses. Progression comes with additional qualifications and experience.
- Non-trade-based careers incorporate architects, quantity surveyors, civil engineers and planners. Many of these professions require higher education, along with practical experience working alongside an established professional. Progression comes through experience, reputation, and the size and scope of projects undertaken.

There are other parts of the sector that have a mix of both of these aspects:

- A training officer or assessor is normally either an experienced tradesperson or professional, with experience in their particular area of work.

- An electrical services inspector is a qualified electrician with a thorough working knowledge of electrical installation, maintenance and repair.

Off-site roles and responsibilities

Off-site roles refer to those who do not spend the bulk of their time on the construction site itself, but remain an important part of the work team. The table below outlines these different roles and responsibilities.

Off-site roles and responsibilities

Role	Responsibilities
Client	The client, such as a local authority, commissions the job. They define the scope of the work and agree on the timescale and schedule of payments.
Customer	For domestic dwellings, the customer may be the same as the client, but for larger projects a customer may be the end user of the building, such as a tenant renting local authority housing or a business renting an office. These individuals are most affected by any work on site. They should be considered and informed with a view to them suffering as little disruption as possible.
Architect	Architects are involved in designing new buildings, extensions and alterations. They work closely with clients and customers to ensure the designs match their needs. They also work closely with other construction professionals, such as surveyors and engineers.
Consultant	Consultants, such as civil engineers, work with clients to plan, manage, design or supervise construction projects. There are many different types of consultant, all with particular specialisms.

Role	Responsibilities
Main contractor	This is the main business or organisation employed to head up the construction work. They organise the on-site building team and pull together all necessary expertise. They manage the whole project, taking full responsibility for its progress and costs.
Clerk of works	This person is employed by the architect on behalf of a client. They oversee the construction work and ensure that it represents the interests of the client and follows agreed specifications and designs.
Quantity surveyor	Quantity surveyors are concerned with building costs. They balance maintaining standards and quality against minimising the costs of any project. They need to make choices in line with Building Regulations. They may work either for the client or for the contractor.
Estimator	Estimators calculate detailed cost breakdowns of work based on specifications provided by the architect and main contractor. They work out the quantity and costs of all building materials, plant required and labour costs.
Sub-contractor	They carry out work on behalf of the main contractor and are usually specialist tradespeople or professionals, such as electricians. Essentially, they provide a service and are contracted to complete their part of the project.
Supplier/ wholesaler contracts manager	They work for materials suppliers or stockists, providing materials that match required specifications. They agree prices and delivery dates.

On-site roles and responsibilities

Just as there is a wide variety of off-site members of the construction team, there are also those who work on site on a daily basis. The following table outlines their different roles and responsibilities.

On-site roles and responsibilities

Role	Responsibilities
Apprentice	They can work for any of the main building services trades under supervision. They only carry out work that has been specifically assigned to them by a trainer, a skilled operative or a supervisor.
Unskilled operative	Also known as labourers, these are entry-level operatives without any formal training. They may be experienced on sites and will take instructions from the foreman or site manager.
Building services engineer	They are involved in the design, installation and maintenance of heating, water, electrics, lighting, gas or communications. They work either for the main contractor or the architect, and give instruction to building services operatives.
Building services operative	This includes all the main trades involved in installation, maintenance and servicing. They take instruction from the building services engineers and work with other individuals, such as the foreman and charge hand.
Charge hand	This person supervises a specific trade, such as plumbing or electrical work.
Foreman	This person supervises the day-to-day running of the site, and organises the charge hand and any building services operatives.
Site manager	This person runs the construction site, makes plans to avoid problems and meet deadlines, and ensures all processes are carried out safely. They communicate directly with the client.
Supervisor	The supervisor works directly for the site manager on larger projects and carries out some of the site manager's duties on their behalf.
Health and Safety officer	This person is responsible for managing the safety of the construction site. They will carry out inspections, provide training and correct hazards.

DID YOU KNOW

A building services engineer may spend as much as 75% of their time in an office and 25% on site, planning and coordinating installations.

Site visitors' roles and responsibilities

There are other individuals who have a role to play on a construction site, but may only make occasional visits, depending on the work being carried out and the demand for their services. The roles and responsibilities of these site visitors are described in the following table.

Site visitors' roles and responsibilities

Role	Responsibility
Training officer and assessor	This person works for an approved training provider. They visit the site to observe and talk to apprentices and their mentors or supervisors. They assess apprentices' competences and assist them in putting together portfolios of evidence.
Building control inspector	This person works for the local authority to ensure that the construction work conforms to regulations, particularly the Building Regulations. They check plans, carry out inspections, issue completion certificates, work with architects and engineers and provide technical knowledge on site.
Water inspector	This person carries out checks of plumbing and drainage systems on construction sites. They will ensure that pipework is properly installed, stopcocks and valves are operable, fittings are fit for purpose, underground pipes are at the right depth, and that there is sufficient protection against **backflow**.
Health and Safety Executive (HSE) inspector	An HSE inspector from the local authority can enter any workplace without giving notice. They will look at the workplace, the activities and the management of health and safety to ensure that the site complies with health and safety laws. They can take action if they find there is a risk to health and safety on site.
Electrical services inspector	Inspectors are approved by the National Inspection Council for Electrical Installation Contracting. They check all electrical installation has been carried out in accordance with legislation, particularly Part P of the Building Regulations.

Documents and documentary procedures

As you progress in your career in building services engineering, you may come across a number of different documents that are used either in the workplace or are provided to customers or clients. All of these documents have a specific purpose. Their exact design may vary from business to business, but the information contained on them will usually be similar.

Backflow: Waste or contaminated water entering the mains or freshwater supply, prevented by means of an air gap in the pipe system.

Documents in the workplace

This group of documents tends to be used only within the workplace. Their general purpose is to collect information or to pass on information from one part of the business to another.

Documents in the workplace and their purpose

Document	Purpose
Job specification	This is a detailed set of requirements that covers the construction, features, materials, finishes and performance specifications required for each major aspect of a project. It may, for example, require a particular level of energy efficiency.
Plan or drawing	This is prepared by an architect. It is drawn to scale and provides a standard detailed drawing. It will be used as a blueprint by building services engineers and operatives while they are working on the site. A typical example would show an entire water system, along with waste and fixtures, and specify the dimensions of all pipework and joints.
Work programme	This is a detailed breakdown of the order in which work needs to be completed, along with an estimate as to how long each stage is likely to take. For example, a certain amount of time will be allocated for site preparation and then piling and the construction of the substructure of the building. The work programme will indicate when particular skills will be needed and for approximately how long.
Purchase order	This document is issued by the buyer to a supplier. It details the type of materials, quantity and the agreed price. The order for materials will have been discussed with the supplier before the purchase order is completed. Many purchase orders are now transmitted electronically, although paper records may be necessary for future reference.
Delivery note	This is issued to the buyer by the supplier. It acts as a checklist for the buyer to ensure that every item requested on the purchase order has been delivered. The buyer will sign the delivery note when they are satisfied with the delivery.
Timesheet	This is completed by those working on site and is verified by the site manager. It details the start and finish times of each individual working on site and forms the basis of the pay calculation for that worker.
Policy document	This covers health and safety, environmental, or customer service issues, among others. It outlines the requirements of all those working on the site. It will identify roles and responsibilities, codes of conduct or practice, and methods and remedies for dealing with problems or breaches of policy.

ACTIVITY

Some larger construction sites have time cards. Each worker has to insert their individual card into a time stamp machine when they start and end their shift.

Why is it important for an employer to record the start and end time of tasks or the hours completed by particular workers?

KEY TERMS

VAT: Value Added Tax is charged on most goods and services. It is charged by businesses or individuals that have raised invoices in excess of £73,000 per year.

Documents for customers and clients

Some documents need to be provided to customers and clients. They are necessary to pass on information and can include records of costs and charges that the customer or client is expected to pay for work carried out. The following table describes what these documents are and their purpose.

Documents for customers and clients and their purpose

Document	Purpose
Quotation and tender	Quotations provide written details of the costs of carrying out a particular job. They are based on the specification or requirements of the customer or client. They will usually be written by the main contractor on larger sites. A tender is usually a sealed quotation submitted by a contractor in the hope that their quotation will not only match the requirements but will also be the cheapest and therefore the most likely to win the work.
Estimate	An estimate differs from a quotation because it is only a guess at what a job may cost. It is not a binding quote but a rough idea of the cost based on what the contractor thinks the work may involve.
Invoice	An invoice is a list of materials or services that have been provided. Each has an itemised cost and the total is shown at the bottom of the document, along with any additional charges such as **VAT**.
Account statement	This is a record of all the transactions (invoices and payments) made by a customer or client over a given period. It matches payments by the customer and client against invoices raised by the supplier. It also notes any money still owing or over-payments that may have been made.

Document	Purpose
Contract	A contract is a legally binding agreement, usually between a contractor and a customer or client, which states the obligations of both parties. A series of promises or agreements are made as part of the contract. It binds both parties to stick to the agreement, which may detail timescales, level of work or costs.
Contract variation	Contract variations are also legally binding. They may be required if both the supplier and the customer or client agrees to change some of the terms of the original contract. This could mean, for example, additional obligations, renegotiating prices or new timescales.
Handover information	Once a project, such as an installation, has been completed, the installer that commissioned the installation will check that it is performing within expected boundaries. Handover information includes: ■ the commissioning document detailing the performance and the checks or inspections that have been made ■ an installation certificate showing that the work has been carried out in accordance with legal requirements and the manufacturer's recommendations.

Types of businesses and companies

When you work in building services engineering, you may work directly or indirectly for a wide range of different types of businesses. You will discover that each business has its own peculiarities and ways of working. It may also be the case that when you are working on domestic contracts, the expectations of the customer or client are entirely different to those of larger commercial contracts. You will also find that many of the normal day-to-day operations are somewhat dissimilar. These differences mean that the methods of communication and documents used can also vary. For example, larger organisations may have more complicated communication systems and require more paperwork to be completed.

Different types of businesses

The first thing to note about different types of businesses is their relative size. A sole trader, for example, may work on their own, running all parts of the business and carrying out the work themselves. At the other end of the scale is a local authority, which may have thousands of employees and also hire many hundreds who have some involvement in building services work. It will have a range of departments, all of which you may have to communicate with at various times to agree contracts, receive payment or gain permission.

DID YOU KNOW

There are more than 165,000 different companies working in the building services industry. The vast majority of them (nearly 95%) employ no more than 13 people.

Sole trader

Sole traders own their own businesses. They carry out all of the tasks themselves, from quotations to buying and organising materials and then doing the work on site. Some sole traders are also employers, so they may have some other building services operatives working directly for them. They will almost always have an apprentice.

It is very easy to set up as a sole trader, but there is no real job security. You must organise your own pension. Also, if you go on holiday or are ill, you will not be earning money.

Many tradespeople like to be sole traders, as there is little in the way of paperwork, they can make their own decisions and, after tax, they can keep all of the profits.

Fig. 5.3 An electrician who works for himself is a sole trader

Contractors

Contractors do not necessarily have to be big companies. They can be an individual or a small number of people with the skills to be able to coordinate and source expertise to put together a quotation for a building project.

Contractors may choose to set up different types of businesses. A small number of individuals may form a partnership. They will create a legal partnership and will usually have complementary skills and experience. Together, their abilities can offer a wide range of services to potential customers and clients.

Fig. 5.4 Contractors may be big or small businesses that coordinate all the work

DID YOU KNOW

To become a limited company, you need to register with Companies House at www.companieshouse.gov.uk. This is the official UK government register of all companies.

Many contractors are limited companies. There are two types:

- Private limited companies (Ltd) have directors and owners who have shares in the company. They are registered companies.
- Public limited companies (plc) are usually larger businesses – many of the main contractors are plcs. Their shares are available for anyone to buy or sell. They are controlled on a day-to-day basis by a board of directors.

The advantage of being a limited company is in the name – 'limited' means that the owners of the business are not personally liable for any debts that the business runs up. They only stand to lose the amount of money that they have put into the company. On the down side, they have to provide detailed accounts and complete much more paperwork than sole traders or partnerships.

Public funded bodies

These are often large service organisations that work on a budget and receive funds directly from the local population (in the case of local authorities) or from central government (in the case of the National Health Service).

Public funded bodies are major clients of the building services industry. They are required to provide certain services to the general public and any money that they spend on building projects must go through a rigorous process of tendering. This is so that the organisation gets the best value for money for any work carried out.

These organisations can be huge, with many departments involved in decision making. If, for example, a contractor was asked to tender for the refurbishment of offices then it would have to deal with both the local authority as a client, and the actual users of the office as customers. In addition, they may deal with the accounts department for payment, a contracts department and a building control department.

Domestic or industrial/commercial contracts

There may be differences when working on domestic or industrial/commercial contracts, as described in the table below.

Working arrangements for domestic and industrial/commercial jobs

Working arrangement	Domestic jobs	Industrial/commercial jobs
Health and safety	The tradesperson needs to manage their own hazards and risks, inform and train their workforce, and cooperate with the client.	A project manager will look at health and safety in the planning phase and will review the situation and set site rules. The contractor must ensure that all workers follow health and safety legislation and carry out their work safely.
Job progress	A foreman or site manager will request regular updates on work progress from others on site and then feed back this information to the client.	There will be a clear line of communication where the main contractor and site manager receive updates from supervisors, foremen, charge hands and building services engineers. They, in turn, will check with others on the site as to progress and any potential delays.
Problems and issues	These will depend on the nature of the problem. The individual coordinating the work may have to talk to the client, architect or consultant, and arrange additional inspections or call in a specialist for advice.	This will also depend on the nature of the problem or issue. Small day-to-day issues will be handled by the building services engineers, charge hands and foremen. More significant problems will be dealt with by the site manager or the main contractor.
Variation or additional work	The customer or client may request a change of agreed work to be carried out. After discussion with other tradespeople, an indication as to any additional time, costs or complications will then be fed back to the client.	This will depend on where the request to make changes has been made. If it is from the client or architect, then the request will filter through the main contractor off site and then the site manager, who will then inform others of the changes. If changes come from on site, then this will have to be cleared by going up the communication ladder so that the changes are approved by the client and main contractor.

Working arrangement	Domestic jobs	Industrial/commercial jobs
Tools, equipment, materials delivery and storage	Smaller jobs may lack storage space and deliveries may have to be dealt with immediately in order not to block roads or access. Tradespeople will usually bring their own tools and equipment and take them away at the end of each working day.	Tools and equipment may be provided by the main contractor and stored in a secure location overnight or while not being used. There will be a dedicated delivery and storage area for materials.
Access	Access to a domestic dwelling will rely on the client providing keys or arranging for someone to be on site at the start of the day.	Steps will need to be taken to prevent unauthorised access to the area of work, usually by fencing or other controls.
Documentary and reporting procedures	Usually this is informal, although the customer will have to be presented with invoices and handover information at the end of the installation.	There is likely to be a lot more paperwork involved, particularly as there are numerous groups of people that will need to be aware of key documents and progress, for example, those related to materials, deliveries, necessary purchases and costs.

Communication and documentary procedures

Although the building services industry, like other sectors, has adopted the use of emails instead of paper documentation and mobile communication instead of formal meetings, the types of communication and documents required may differ from job to job. They may also be dependent on the type of business or organisation involved, which may have its own procedures.

The following table provides some suggestions as to the variations in communication and documentary procedures for different types of business.

REMEMBER

A stamped, dated and signed document is often required to prove work has been carried out to particular standards and has been inspected.

Ways different types of business communicate

Type of business	Communication and documentary procedures
Sole trader	■ Face-to-face and telephone communication is usually preferred. ■ Basic documentation, such as invoices, contracts and estimates are necessary for some jobs.
Contractor	These will depend on the size of the organisation and the preferred ways of carrying out work. ■ Formal meetings may be necessary both on and off site. ■ The contractor may want regular updates in writing. ■ They will require paper evidence of any costs and to match purchase orders to delivery notes and invoices. ■ They will require signed contracts, commissioning documents and installation certificates.
Public funded body	Usually this is the most demanding in terms of communication and documentary procedures. ■ They will require face-to-face update meetings on a regular basis. ■ Regular site visits will be made. ■ Paper copies of all relevant documents will be needed, including invoices, contracts, commissioning documents and installation certificates. They may also require multiple copies of documents to be sent to various parts of the organisation.

DID YOU KNOW

A principal contractor has to be appointed on construction projects that will last for more than 30 days or 500 man days.

KEY TERMS

Principal contractor: This is either an individual working for 30 days or more on a construction job or 50 people working for 10 days or more.

CHECK YOUR KNOWLEDGE

LEVEL 2

1. What do you understand by the term sustainability as applied to buildings?
2. What is a blueprint?
3. What kind of work would an apprentice be involved in?
4. Name four practical applications for building environmental management systems.
5. What are the main differences between a direct and an indirect cold water system?
6. Where might you find a soil pipe?
7. What is the SI unit of measurement of electric current?
8. From where might a dwelling produce greywater?
9. What is the role and responsibility of a clerk of works?
10. Who would a building control inspector work for and what would they be looking at?
11. What is a purchase order, who creates it and who is it sent to?
12. What is the difference between a quotation and a tender?
13. A tradesperson who works alone, organising all work and taking full responsibility, is likely to be operating what type of business?
14. Give an example of a public funded body.
15. How might reporting job progress on a domestic job, as opposed to an industrial or commercial job, differ?

6 Refrigeration and air conditioning operations

This chapter covers the learning outcomes for:

Understand and demonstrate fundamental refrigeration and air conditioning (RAC) operations

City & Guilds unit number 105; EAL unit code QACC1/05; ABC unit code A05 for Level 1 and A03 for Level 2

Refrigeration and air conditioning (RAC) systems are used either to cool a dwelling or building's environment or to cool products. Essentially, both systems transfer heat away from a cool, low-energy reservoir into a warm, higher-energy reservoir.

Refrigeration systems are designed for domestic purposes, such as refrigerators, and for industrial processes, such as chilling plants.

Air conditioning systems have a number of different combinations. These include:

- air conditioning for machines or spaces
- split air conditioners
- units for larger buildings.

IN THIS CHAPTER YOU WILL LEARN ABOUT:

- RAC health and safety procedures
- RAC hand tools
- RAC materials and components
- basic RAC practical applications.

RAC health and safety procedures

Working with refrigeration and air conditioning means that some extra precautions are necessary. Above all, you need to make sure that you exercise care, use appropriate personal protective equipment (PPE), and make every attempt to reduce risks.

PPE

When handling refrigerants, you will need to wear PPE that protects your eyes and hands.

REMEMBER

PPE should always be worn when handling refrigerants, for additional protection.

Eye protection

It is important to wear approved mechanical safety goggles when you are:

- moving, connecting or installing cylinders
- working on piping and fittings containing refrigerants
- charging and recovering.

Fig. 6.1 Protective gloves should always be worn when operating compressors

Hand protection

You should always wear thermal protection gloves. This is because compressors can be extremely hot and could cause skin burns. Also, some parts of the system may be at sub-zero temperatures, which could cause frostbite on contact.

Health and safety practices

It is the responsibility of the employer (or the independent engineer) to take all reasonable steps to protect the health and safety of all those working on refrigeration and air conditioning systems. They also need to ensure the safety of contractors, clients and any others who might come into contact with RAC systems while the work is being carried out.

The employer will 'have the responsibility, as far as reasonably practicable'. In practice this means that they must ensure:

- the safe provision and maintenance of plant and systems of work
- the absence of risks to health when using, handling, storing and transporting articles and substances
- that instructions, information, training and supervision are provided, as necessary, to ensure the health and safety of all employees
- the provision and maintenance of a safe place of work and safe access to and departure from that place of work
- the provision of a working environment that is safe and without risks to health
- adequate welfare facilities and arrangements
- compliance with all relevant statutory requirements for health, safety and welfare.

It is vital that a company's health and safety policy is fully implemented, kept up to date and formally reviewed at least once a year. RAC companies also need to make sure that all of their employees are aware of their responsibilities under health and safety laws – that is, to work safely and to cooperate in the maintenance of safe working conditions.

Workshop hazards and risk reduction

Working with refrigerants can be hazardous. The table below outlines the types of hazards you may experience when working with RAC systems and how you can try to eliminate them.

Hazards when working with RAC systems and the necessary precautions to take

Hazard	Precautions
Ignition sources	■ Eliminate open flames or other ignition sources such as operating thermostats ■ Do not light halogen flame detectors, candles or matches ■ Remove or eliminate hot surfaces from near the refrigerant or work area, including bare light bulbs
Sparks and electrocution	■ Replace or install shields over electrical contacts ■ Fix any loose capacitors or electrical ignition sources
Disconnection	■ Put a warning tag on electrical disconnect switches to protect from unexpected start-ups ■ Take out the fuses or circuit breakers before working
Energy in compressed gases	■ Relieve the system's gas pressure before working ■ Pump out and recover the refrigerant

Reporting procedures

All organisations have different procedures in place for reporting workshop hazards and reducing risks. It will be your responsibility to report immediately to another member of staff any situation that you have reason to believe will present a hazard or health risk, including equipment malfunctions.

Basic electrical safety checks

- Ensure that all sources of electrical ignition have been removed.
- Check that electrical equipment and lines are in good condition.
- Isolate electricity from the work area.

Basic mechanical safety checks

- Make sure any guards or shields are installed and are in good condition.
- Look for any trapped wires.
- Secure any loose clothing.

Basic health and safety checks

- Ensure that you have adequate eye protection and other PPE.
- Make sure you know the location of the fire extinguishers and which ones to use in the event of a fire (see page 62).
- Check that there is a clear exit route from the work area, should you need to evacuate.

Refrigeration and air conditioning systems

Refrigeration and air conditioning systems are used to cool buildings, machines and perishable goods, such as food stock.

The most common system is the vapour-compression refrigeration system. This uses chemical liquids, called refrigerants, which have very low boiling points. When pressurised and forced to be a liquid they give out heat. When they evaporate they cool what they are in contact with because they absorb heat.

A description of how the vapour-compression refrigeration system works is given on pages 161–2.

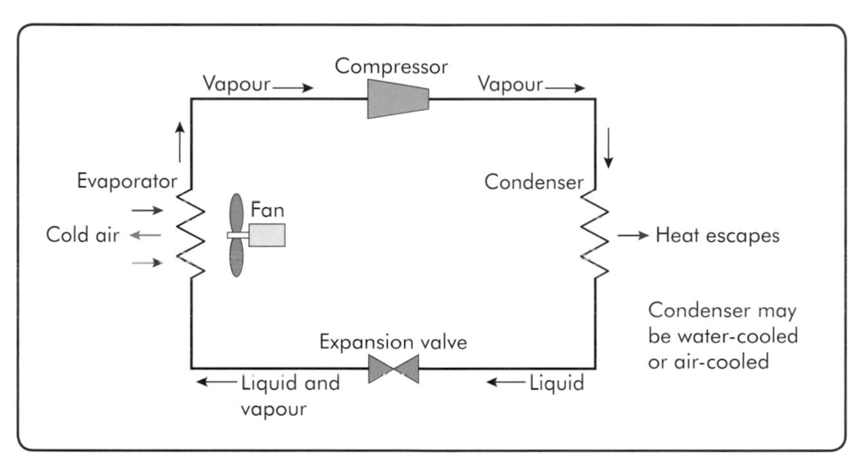

Fig. 6.2 A simple vapour-compression refrigeration system

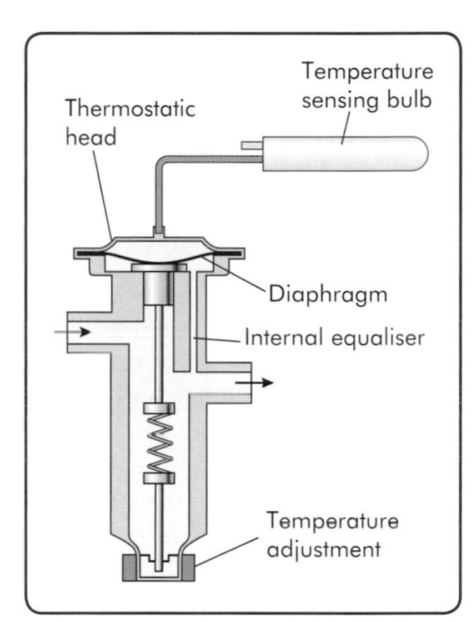

Fig. 6.3 The thermal expansion valve

Refrigerants in a vapour-compression system move round a sealed circuit. They move the heat from one part of the circuit to another.

RAC hand tools

Many of the tools used in the RAC part of the building services industry are common to others, such as plumbers.

Some of the hand tools used in RAC are listed in the table below, however this is not an exhaustive list. The more complex the job and the more demanding the systems, the broader the range of tools you will need.

Key RAC hand tools, their use and maintenance

Tool	Use	Maintenance
Adjustable spanner	To loosen or fasten nuts or bolts. Always have the movable jaw in the direction you wish to rotate. This is more efficient and prevents damage or slipping.	Replace if the teeth are worn or the ratchet slips. Keep clean as jointing compounds could cause it to slip.
Screwdriver (crosshead and flat)	Always check for the best fit in terms of diameter and depth. Avoid using the wrong screwdriver (e.g. flat head for crosshead screws)	Discard if the handle is broken. Flat screwdriver tips that become misshapen can be filed down if necessary.
Junior hacksaw	Always make sure that the right blade is used for the material. For example, a 24-teeth per inch blade should be used for cutting plastics and copper, and a 32-teeth per inch blade should be used for low carbon steel pipes.	Replace blades that have defective or worn teeth. Always check the blade is tightened. Make sure the teeth point away from a forward cut.
Pipe cutters	These are used for making straight, precision cuts in lengths of pipe. Make sure the right blade is used for the material.	Always replace blunt or damaged cutter wheels. Lubricate the wheel and rollers.
Deburr tool	This is used for to remove undesirable sharp edges, burrs and fins remaining on a cut piece of pipework.	Ensure the blade is kept sharp and that it is not deformed.
Flaring/ swaging tools	These are used to widen the end of a length of pipe, so that a second length can be added to provide a more secure fit. Ideally, use one that converts quickly with a screw feed.	Check cones for damage and keep the vice well lubricated.
Brazing equipment	Brazing is joining two metals with a filler metal. Only the brazing filler metal is melted, not the base metal.	You will need to regularly clean, lubricate, maintain and adjust the equipment to ensure its efficient operation.
Tape measure	This is used to measure distance in inches or centimetres.	Replace if it is snapped or sheared, has sharp edges or fails to retract after use.
Level	This is used for fixing to the correct level (square). Ideally, you should have a small and a large level.	Ensure that the levels are not broken or dirty.

Other tools that you may need to use include:

- a blow torch
- a compact tube cutter (ideal for getting into tight spaces)
- a set of files (for working on plastic piping for push-fit connections)
- a high-speed drill (with set of drill bits for different materials)
- insulated screwdrivers (when working with electrical connections)
- a retractable utility knife (for general purpose trimming)
- a pipe wrench (for tightening up steelwork).

You will need to be able to show that you can use a range of hand tools to carry out basic RAC tasks safely. You will also need to show that you understand how to care for and maintain RAC hand tools.

Try to practise as much as you can with your hand tools. Get into the habit of ensuring that they are clean and undamaged, and replace any defective tools promptly.

TRADE TIP

It can be a false economy to buy cheap hand tools, as the metal parts are often too soft or lack the strength for regular use.

ACTIVITY

Choose three tools from the list on this page. Examine them and write down the things you are looking for to make sure they are in a good enough condition to use.

RAC materials and components

When working on RAC systems, you will routinely use a wide variety of materials. However, copper is the preferred material for use with all refrigerants (except ammonia), so it is one of the most common materials you will use.

TRADE TIP

IET has an *Electricians' Guide to the Building Regulations*. This is a good starting point for any domestic installer.

Fig. 6.3 Copper piping is used regularly in RAC

Sizes and types of copper piping

Copper piping is the ideal material for most RAC applications because of its strength, the ease with which it is fabricated and soldered, and its high conductivity to heat. It is used in the fabrication of RAC equipment and for the installation of systems.

Copper piping comes in various sizes and types. Design documentation states tube sizes and wall thickness, and manufacturers mark the external surface of tubing to show that it conforms to BS EN 378:2008, which is the most critical standard for refrigeration and air conditioning system safety, design and commissioning. The copper tube also has to conform to BS EN 12735-1:2010, which means that the copper is refrigeration and air conditioning grade.

Alloy: A combination of two or more metals, which is designed to have greater strength or resistance to corrosion.

Anneal: A heating and cooling process that alters a material's properties.

Copper piping is normally available in a variety of sizes, as the table below shows. Note that the types of copper, known as copper **alloys**, are distinguished by temper levels. These indicate the way in which the metal has been cold worked and then **annealed** to give a certain degree of flexibility. Typical tempers for copper alloys are soft, half-hard (or medium), and hard.

Types and sizes of copper piping

Type of pipe	Size of pipe	Size of lengths or coils
Soft drawn (table W and Y)	⅛ inch; 3⁄16 inch; ¼ inch; ⅜ inch; ½ inch; ⅝ inch; ¾ inch; ⅞ inch	6 m and 15 m coils
Half-hard temper (table X)	⅜ inch; ½ inch; ⅝ inch; ¾ inch; ⅞ inch; 1⅛ inch; 1⅜ inch; 1⅝ inch; 2⅛ inch	3 m and 6 m lengths
Hard temper (table Z)	2⅝ inch; 3⅛ inch; 3⅝ inch; 4⅛ inch	3 m and 6 m lengths

Fittings

Fittings are designed to join onto copper tubes and become part of the pipework system itself.

There is a variety of fittings for copper tubes, which can be purchased commercially. These include:

- flare fittings
- compression fittings
- mechanical couplings
- pipe flanges.

In addition, there are many fittings that are specifically manufactured for the industry.

TOOLBOX TALK

All refrigeration grade copper tubing should be clean, dehydrated and sealed at both ends.

Elbows

Elbows are basically designed to allow a change in direction of the piping. They can be:

- long or short radius elbows
- reducing elbows, which are used to fix together two pieces of tube of differing size
- 45° elbows.

Long radius elbows have a centre-to-face dimension of 1.5 × diameter. They are more common and are used when flow is more critical and when there is available space.

Short radius elbows have a centre-to-face dimension of 1.0 × diameter. They are used in situations where the pipe is being fitted in a tight area.

Reducing elbows are used to fix together two pieces of tube of differing size.

Allow completed brazed joints to cool naturally. Cooling with water gives a weaker joint and if it enters the pipework it can damage the final system.

Tees and unions

Tees are used to either combine or split fluid flow, and can be equal or reducing pieces. The most common are those with the same inlet and outlet sizes.

A union allows for the easy disconnection of pipes for maintenance. They are often used instead of couplings, which permanently connect two pipes together.

TOOLBOX TALK

Always wear leather gloves when handling newly brazed copper pipework. Copper is a very good conductor of heat and can retain that heat for a considerable time after brazing is finished.

Jointing methods – brazed and flared

Brazing is very similar to soldering, but uses much harder materials at higher temperatures. Brazing uses heated metal filler. **Flux** is used to prevent oxides from forming when the metal is heated. Oxides would weaken the bond.

Always ensure adequate ventilation before brazing commences. A 'hot work' permit will be required before brazing on site is normally allowed.

A flaring tool is used to enlarge the tubing so that it matches the tapered end of the flare fitting. A flare nut is installed over the tubing and is then tightened. Flare connections are very reliable, but can be labour intensive. They are very secure against leaks.

KEY TERMS

Flux: A substance that is used to prevent oxidation during soldering.

Superheat: When the temperature rises above boiling point. Superheated substances are in a vapour state.

Thermal expansion valve

The thermal expansion valve (TXV or TEV) is used for refrigerant flow control. It operates at varying pressures resulting from varying temperatures.

The purpose of the valve is to maintain constant **superheat** in the air conditioning evaporator. The valve needs a capillary tube and thermal (temperature-sensing) bulb (or element) to work: the capillary tube connects the element to the top of the TXV diaphragm; the thermal bulb is part-filled with refrigerant.

The valve acts as an expansion device for the refrigerant. It allows the refrigerant to move from the warmer and higher pressure area of the condenser output to the colder and lower pressure area of the evaporator input. The valve achieves this by restricting the flow (in almost the same way as a water tap works when it is nearly shut off). In effect, it means that the refrigerant is at high pressure at the valve input and at low pressure at the output.

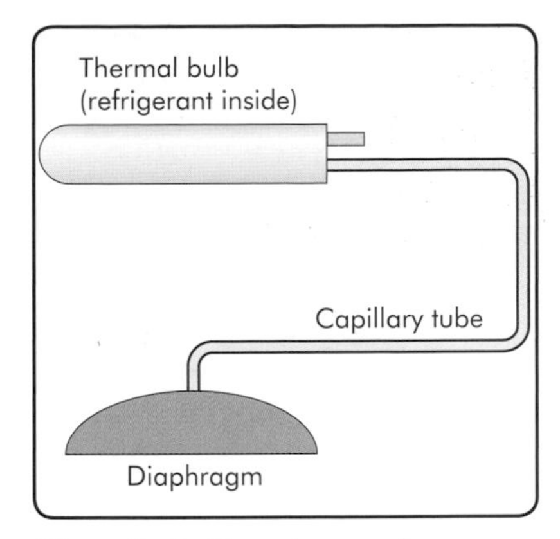

Fig. 6.4 The thermal expansion valve helps maintain constant superheat.

The valve also regulates the amount of refrigerant that flows through the evaporator. The refrigerant at the evaporator output has to be in gaseous form, so the valve keeps the refrigerant just above the vaporisation temperature. It regulates the temperature by opening and closing in response to the thermal bulb temperature. When the temperature rises above the set point, the valve opens up wider and more refrigerant can flow. When it falls below the set point, the valve closes up and allows less of the refrigerant to flow.

ACTIVITY

See if you can label the diagram of a thermal expansion valve shown here.

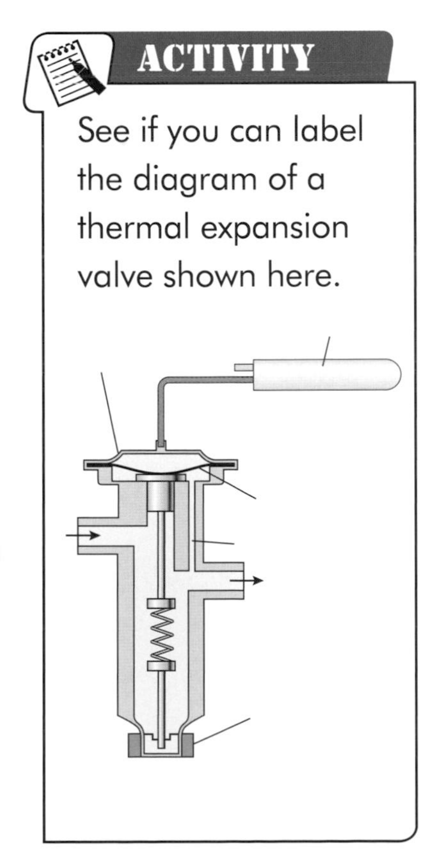

Condensing and evaporating units

All condensing and evaporating units work in the following way:

1. A low-pressure cold liquid refrigerant is supplied to a vessel called an evaporator.
2. In the evaporator, air is forced over the pipes containing refrigerant, usually by means of fans.
3. The refrigerant absorbs the heat from the air, so cooling the air and warming the refrigerant.

Fig. 6.5 A domestic refrigerator

4. The liquid refrigerant, having absorbed the heat from the air, now becomes a vapour.
5. This vapour is drawn into a compressor, where its pressure is raised. In this way, the low-pressure, low-temperature vapour becomes a high-pressure, high-temperature vapour.
6. The vapour now passes from the compressor into another vessel, the condenser. Here, the vapour is cooled using a readily available cooling medium, such as air or water. During this cooling process, the high-pressure vapour returns once again to liquid.
7. The liquid refrigerant now returns to the evaporator via the expansion valve. The expansion valve forces through the high-pressure liquid, rapidly dropping its pressure and temperature to that required in the evaporator.

The cooling process can then begin again.

This cycle of evaporation – compression – condensation – expansion continues until the desired temperature is achieved.

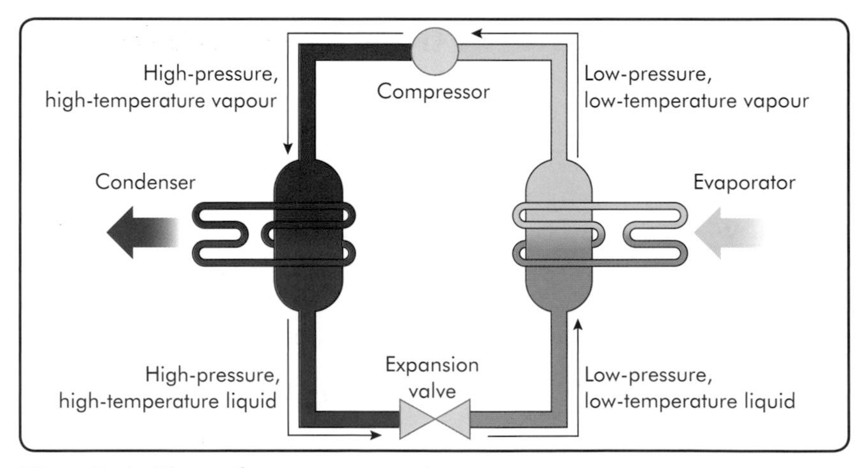

Fig. 6.6 The refrigeration cycle

Clips

Clips are used to hold the connectors in place before they are inserted into the hose. This is done by sliding them over usually flexible pipes. The clips can be tightened using a special pair of pliers. The use of the clips ensures that the connector does not work itself loose through handling or when the system is up and running.

Fixing devices

The other key issue is providing support for piping. To achieve this, the pipes need to be fixed with support or hangers that are capable of carrying the weight of the pipes, including the contents (and in some cases the insulation as well). The pipework may have to be routed through floor ducts and, in other cases, it will need fastening to walls or ceilings.

The process of fixing to walls or ceilings offers different fixing opportunities:

- Pipe clips are pipe saddles that clamp over the pipes and are then securely fixed to the wall or ceiling.
- Cable trays are ideal if there are a number of pipes that need to be run. The tray is fixed to the wall or ceiling with a spacer and then the pipework is fixed to the tray using nylon ties.
- Other fixing systems can include clamps and support brackets that are designed to fix either single or multiple pipes.

It is always important to ensure that the correct type of screw and length is used in all cases. Screws must be suitable for the material into which they are being screwed, and they need to be strong enough to support the weight of the copper tubing.

Refrigeration and air conditioning systems use slighter more complex versions of the basic refrigeration system. The fundamentals remain the same so, although each different manufacturer of a system has a slightly different setup, there will always be key components and processes that are familiar to you.

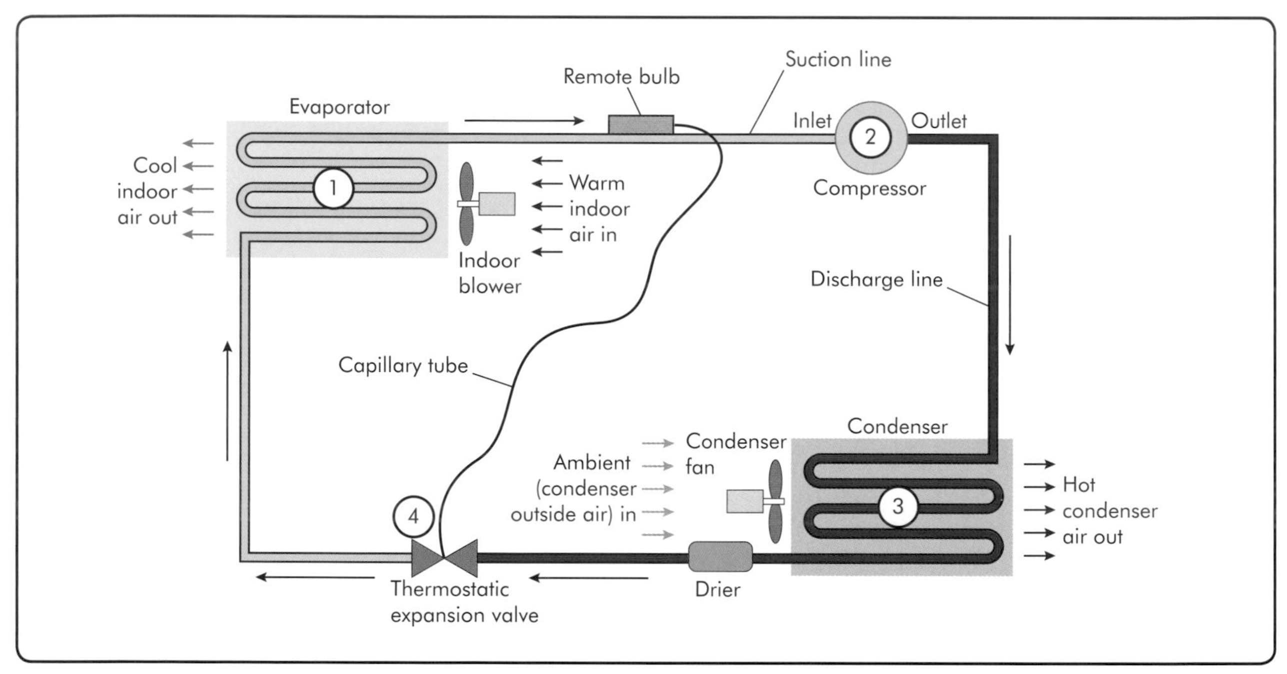

Fig. 6.7 An air conditioning unit is a more complex system. Parts 2, 3 and 4 are usually situated on the outside of a building

Cleaning materials – flux and rods

It is important to make sure that the copper pipe remains clean and that no objects or obstructions are left in the piping. A wide variety of cleaning materials can be used, including:

- Pen reamers to remove burrs – they are pen-sized tools with single blades
- Pipe cleaning tools have brushes in order to clean and prepare piping and fittings – they aim to eliminate any scrapes or rough edges
- Cleaning brushes are stiff steel bristles for cleaning out copper fittings and sockets or joints prior to soldering.

The flux has to be chemically compatible with both the filler metal and the base metal.

Flux

Flux is used to clean off the brazing surface (see page 160). Flux can be a paste, a liquid or a powder. It can also be coated onto the brazing rods.

The flux flows into the joint and is then displaced when the molten filler metal enters the joint. It is important to make sure that any excess flux is removed, as this could lead to corrosion.

Soldering and brazing rods

Soldering irons are hand tools that are used to melt solder to flow into a joint between two pieces of metal. They have a heated metal tip and an insulated handle.

In many situations in RAC, brazing is recommended. The technique is very similar to soldering, but takes place at much higher temperatures and uses brazing rods not solder. In most cases, torch brazing will be used on site. This is ideal, as it allows the engineer to work on specific joints. The temperatures used to melt the filler are above 450 °C.

Basic RAC practical applications

Drawings and specifications

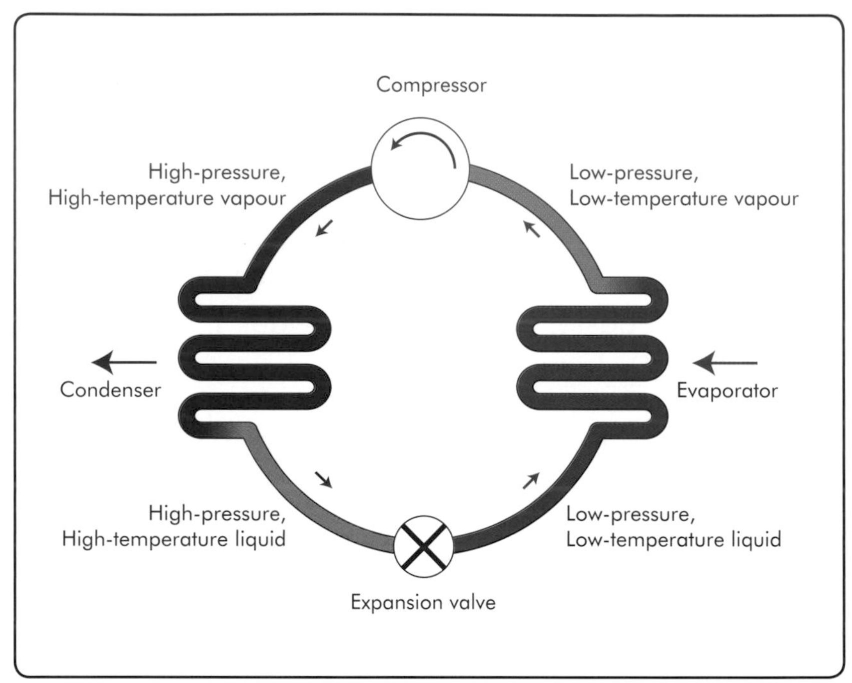

Fig. 6.8 A simple refrigeration circuit

The simple circuit shown in Fig. 6.8 consists of the four basic components:

- evaporator
- compressor
- condenser
- expansion device.

Clearly, all of these basic components need to be connected together. You will be provided with a suitable pipework assembly diagram. This will enable you to:

- measure and cut pipework to length
- bend and braze pipework
- pressure test the system.

Industry procedures

There are various tried and tested ways of making sure that pipework is completed in the correct manner. Over time you will learn all of these techniques and understand why it is important to follow specific procedures, which have been tried and tested over the years.

Measuring and marking out for pipework

Measuring and marking out is the process of transferring a design or specification onto the actual site. It is the first stage of the installation process. Many labour-saving tools can be used to ensure correct measurements are made and that you minimise the number of additional connections or bends needed within the system.

You need to be able to interpret drawings and pipe details. To do this you will need:

- a pencil or marker
- a spirit level
- a folding steel rule
- a tape measure
- the drawing.

Once the measuring and marking out of the pipework has been completed, you will be able to see exactly what is required in terms of making the necessary bends in the pipework.

TOOLBOX TALK

It is better to use a pipe cutter when cutting refrigeration pipe as hacksaws produce more swarf.

Swarf: Small fragments of a metal produced when it is cut or filed.

REMEMBER

To avoid mistakes, measure twice and cut once.

Cutting pipework

Hacksaws or tube cutters are used for cutting copper pipes. Where possible, it is preferable to use tube cutters rather than hacksaws to reduce **swarf**. The tubing must be cut squarely and, after cutting, the ends of the pipe must be reamed and scraped. This is to ensure that there are no sharp burrs.

It is also important to make sure that no filings or chips get into the pipework. It is also good practice not to leave unconnected and exposed pipework ends unsealed.

Bending copper pipe – 90° bends and offsets

It is a far neater and more professional installation if bends rather than fittings are used. This also reduces the possibility of pressure drops.

Bending pipes is something that needs skill. The normal procedure is as follows:

1. Accurately measure the distance between the centre of the joint and the position where the outside bend is going to finish.
2. ou can bend either by hand, by using an internal spring inside the pipe to stop it being crushed flat, or by machine.
3. A 90° bend requires you to take the measurement from a fixed point (usually the end of a piece of pipe) to the back of the bend. The pencil mark is made from the fixed point to the required length. If the tube is then slotted into the bending machine, it is set up with the mark squared off from the outside of the tube former.

4. The roller is then adjusted and the bend is pulled.

For an offset, a bend that is other than 90° is made. This will allow the pipework to operate around obstacles. The procedure is as follows:

1. Make a first bend in the pipe. This does not need to be completely accurate and will depend on the profile of the object that you are trying to work around.
2. he straight edge of the pipe is now placed against the former.
3. The measurement for the offset is then taken from the inside of the tube, extending to the inside edge of the straight section of pipe.
4. The tube is then adjusted in a machine and the pipe can be bent so that it is in line with the first bend.

As copper pipe used in plumbing uses the same bending methods, see pages 269–71 on pipe bending in the plumbing chapter for more details.

Fabricating pipework

This means cutting, bending and brazing pipework to create the required layout. It is important to remember that copper refrigeration piping is always specified by its outside diameter. In plumbing, the pipework is specified by its inside diameter.

Fabricating pipework for a refrigeration system will require a suitable pipework assembly diagram. This is so that each section can be measured and cut. Any bends can then be created and sections of pipework can be brazed. The final task is to pressure test the system.

Remember that the thickness of a copper pipe wall is around $\frac{1}{16}$ inch. You can then convert between plumbing and RAC pipe diameters.

ACTIVITY

1. What would be the copper plumbing pipe diameter specification if you were using some spare 1⅛ inch refrigeration pipe for a plumbing task?
2. Find out why you would not use copper plumbing pipe for a refrigeration task.

Jointing copper pipe

There are several different ways in which copper tube can be jointed.

Brazing

Brazing provides strong, leak-free joints. The best way of achieving this is by using an acetylene oxygen mix that produces a flame of over 600°C.

1 prepare the joint by swaging one of the ends of the pipe (or using a union).
2. Thoroughly clean and degrease the piping.
3. Perform a test fit – the joints should now be supported.
4. Apply the flux for the brazing alloy.
5. Heat the joint to the recommended temperature, moving the torch around in a circular motion.
6. Apply the brazing alloy to the heated joint. Note that the alloy is not melted by the torch itself.
7. Allow the joint to cool.
8. Clean the joint and remove all flux.
9. Pass oxygen-free nitrogen through the pipework during the operation, to reduce the build up of scale on the inside.

Compression joints

The procedure for a flared or manipulative compression joint is as follows:

1. Cut the tube to the required length and deburr.
2. Place the nut and compression ring onto the pipe.
3. Use a flaring tool to flare out the end of the tube.

4. Assemble the fitting with the angled side of the adaptor fitted into the flared end.
5. Finger tighten the joint before fully tightening it with a spanner or grips.

A non-manipulative joint does not require the pipe end to be flared. Instead, an olive or compression ring is used to provide a watertight seal on the pipe end.

Soldered joints

Although soldered joints are not usually suitable for RAC, this is a similar procedure to brazing joints:

- Cut square, deburr and clean the tubes.
- Apply the flux and assemble the joint.
- Apply heat first to the fitting and then the solder.

Capillary fittings

There are a range of different types of capillary fittings, which include elbows, adaptors, couplings, connectors, tees and offsets.

Pressure testing pipework and fittings

For RAC operations, oxygen-free nitrogen is used for pressure testing. It is introduced to the system using a hose. Pressure gauges are used to read the system pressure and, at the same time, leak tightness testing is also carried out but at lower pressures.

This process will take place as an integral part of the installation programme and at the earliest possible stage. It tests for the overall mechanical strength and leak tightness. All of the components will be tested in this way.

ACTIVITY

Find out what PPE you need and the other safety precautions that must be taken before you pressure test RAC pipework.

CHECK YOUR KNOWLEDGE

LEVEL 1

1. Outline the key pieces of PPE that you need for basic RAC practical applications.
2. How would you go about reporting any potential workshop hazards?
3. What is the purpose of a deburr tool?
4. What are the main sizes of copper pipe used for RAC installation?
5. What is the purpose of a condenser?
6. Why is it important to use the right flux for soldering or brazing?
7. Identify the three ways in which you could bend a copper pipe.
8. What is a flared joint?

7 Electrical installation operations

This chapter covers the learning outcomes for:

Understand and demonstrate fundamental electrical installation operations

City & Guilds unit number 106; EAL unit code QACC1/06; ABC unit code A04 for Level 1 and A02 for Level 2.

In this chapter, you will learn about basic electrical practical applications. You will gain knowledge of the types of tools, materials and equipment used within the electrical industry. You will also consider how to use them safely in accordance with key health and safety legislation.

The electrical industry is a large and diverse sector. It covers domestic, commercial and industrial installations, from domestic lighting and power to industrial heating and specialist installations, such as intruder alarms and CCTV. It also includes electrical repair and maintenance. This means that different organisations require specialist skills, tools, materials and equipment. For example, you will require different expertise for domestic lighting compared to the skills required by an alarm engineer.

IN THIS CHAPTER YOU WILL LEARN ABOUT:

- health and safety when carrying out basic electrical practical applications
- electrical hand tools and their safe use and maintenance
- materials and components used within the electrical industry
- basic electrical practical applications.

Introduction to electrical circuits

The two main types of electrical circuit that you need to understand are:

- the power circuit – otherwise known as the ring main or radial
- the lighting circuit.

KEY TERMS

Junction box: Contains terminals for joining electrical cables.

Power circuits

This type of circuit feeds a 13 A (ampere or amp) socket. The ring main flows in two directions to reach the socket, as shown in Fig. 7.1. **Junction boxes** can be used to extend the circuit.

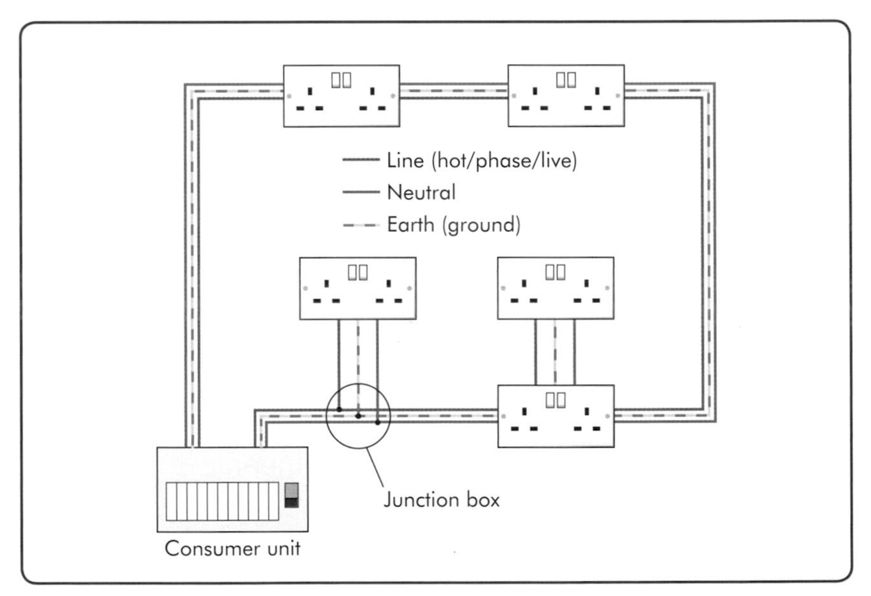

Fig. 7.1 A ring main circuit with a junction box spur and a spur from a socket

TOOLBOX TALK

Part P of the Building Regulations states that only a competent person is allowed to carry out installation and modification of electrical wiring.

The consumer unit will have devices, such as fuses, that control and protect the circuit. Both the power and lighting circuits use **PVC**-insulated cable and are supplied by the consumer unit.

KEY TERMS

PVC: Polyvinyl chloride.

Lighting circuits

A lighting circuit is protected by a 5A or 6A protective device in the consumer unit. The cable is routed from the consumer unit through a number of ceiling roses or wall light fittings. Each of these is, in turn, connected via a light switch.

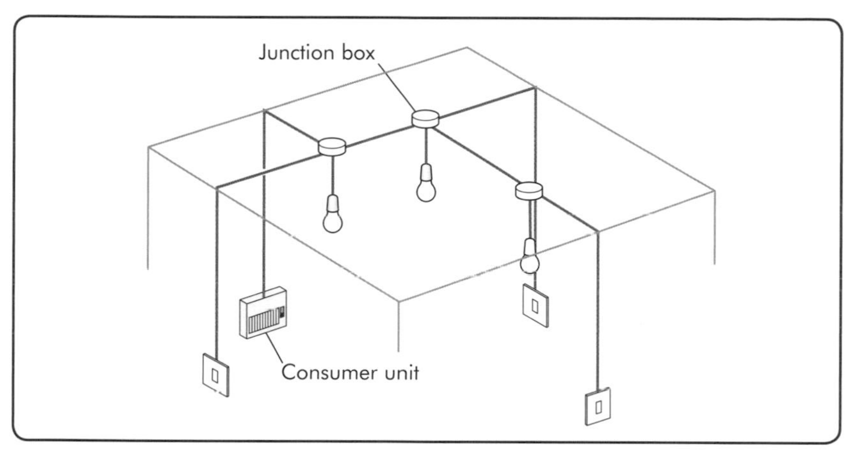

Fig. 7.2 A simple lighting circuit

Health and safety when carrying out basic electrical practical applications

Personal protective equipment

When you work with electrical applications, it is essential that you wear the correct personal protective equipment (PPE). For example, you will need to ensure that you have adequate protection for your:

- eyes (e.g. safety goggles)
- hands (e.g. gloves)
- head (e.g. a hard hat)
- clothing (e.g. overalls)
- feet (e.g. steel-capped boots with rubber soles).

Turn back to Chapter 2, pages 34–6, to remind yourself of the different types of PPE and their uses.

KEY TERMS

Risk: The chance, high or low, that somebody will be harmed by a hazard.

Hazard: Anything that could cause harm.

Identify and report any potential workshop hazards

When carrying out a **risk** assessment, you need to:

- state the task
- identify any **hazards** associated with that task
- name any people who could be affected by those hazards
- account for any existing control measures
- evaluate the risks by considering both the likelihood of the identified hazards causing injury and the possible severity of such injuries
- decide upon any further actions required to reduce the risks, and prioritise those actions.

TOOLBOX TALK

You need to review your risk assessment annually, or when making changes in the workplace, such as new machinery, substances, procedures or employees, or if there is an accident or near miss.

Demonstrate how to reduce the risks for workshop activities

To control any risk of an accident occurring in the workplace, you need to consider the following risk controls. These can be summarised using the acronym ERIC.PD:

- Elimination – it might be possible after a risk assessment to completely remove the hazard and, therefore, any danger associated with it.
- Replacing or substitution – this means replacing items or equipment with safer alternatives. For example, using an electric screwdriver insulated to 1000V in place of a standard screwdriver will help prevent electric shock when working on installations.

- Isolation or enclosure – this is also known as encircling. For example, it could be restricting access to electrical panels or using barriers to prevent entry to working areas.
- Control – this means controlling the working environment by ensuring protective measures and safe systems of work, and informing others of these. One example is a permit to work on live electrical circuits that includes a safe isolation procedure.
- PPE– always use the correct PPE for the work activity.
- Development – this refers to the need for proper supervision and the provision of information and training regarding the workplace and the hazards involved.

DID YOU KNOW

In 2009–2010:
147 workers were killed at work.
121,430 other injuries to employees were reported.
1.3 million people who worked during the previous year were suffering from a work-related illness.
(Source: www.hse.gov.uk/statistics/index.htm**)**

ACTIVITY

Consider what you learnt about health and safety in Chapter 2.

Then identify any hazards in your working environment, for example, chemicals, asbestos, electricity, tools and ladders.

Now assess the risks arising from these hazards. You need to consider all people in the workplace, whether young or old, people with disabilities, apprentices, trainees or visitors.

When evaluating the risks, you need to take into account any existing control measures, such as machine guards or competent persons who are trained in the work activity.

Electrical hand tools and their safe use and maintenance

There are two basic categories of hand tool:

- General engineering, construction work, repair and maintenance hand tools
- Electricians hand tools.

Electricians need specialist tools and equipment due to the nature of their working environment and the dangers associated with electricity.

One of the key differences with tools used by electricians is that they are insulated to provide protection from electric shock. Electricians' tools are insulated to 1000 V, which is classed as low voltage in the BS 7671 IET Wiring Regulations (17th Edition, 2011).

REMEMBER

Only use tools insulated to 1000 V when working on electrical installations.

Electric shock occurs when a person becomes part of the circuit. The severity of shock will depend on the level of current and the length of time it is in contact with the body, but the lethal level is usually 100–200 mA (0.1–0.2 A). At this level of current, the heart fibrillates and breathing stops.

Do not use electrical screwdrivers on live equipment if there is any damage to the tip or insulation.

Screwdrivers

There are various types of screwdrivers, for example:

- a slotted flat tip (engineer's screwdriver) is used for installing accessories and general use
- crosshead screwdrivers (Phillips and Pozidriv) give better torque when used with crosshead screws,

which are often used to secure accessories and terminals in miniature circuit breakers

- electrical screwdrivers are designed and certified by **VDE** for use around live equipment up to 10,000V. They are available in slotted or crosshead.

When using a screwdriver:

- always make sure that the tip fits snugly into the head of the screw as this will prevent damage to the screwdriver
- check that it is dry after use, as moisture left on the screwdriver will cause it to rust and corrode
- replace it if it is damaged, as using it may cause you to slip and harm yourself or damage the work
- only use it for the job it is designed for – screwdrivers are NOT designed for prising open paint cans, chiselling brickwork, hammering, door wedges and lifters, prying or lifting boards.

KEY TERMS

VDE: Verband der Elektrotechnik, the Association for Electrical, Electronic and Information Technologies.

TOOLBOX TALK

Flat tip screwdrivers should never be used on crosshead screws, or as a centre punch or scribe to make holes for screws.

Pliers and cutters

- Side cutters – a key tool for all electricians, they are used every day to cut and prepare wire for connection. Side cutters should not be used to cut cable wider than 2.5 mm^2 as this will damage the cutting blades.

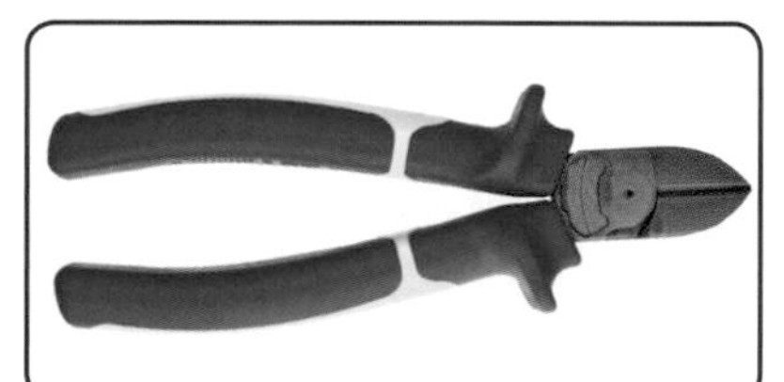

Fig. 7.3 Insulated side cutter

- Combination pliers – as their name implies, they have a combination of features designed to make them versatile. The flat serrated jaw is used to twist

TRADE TIP

Side cutters and pliers should be kept dry and lubricated as per the manufacturer's instructions.

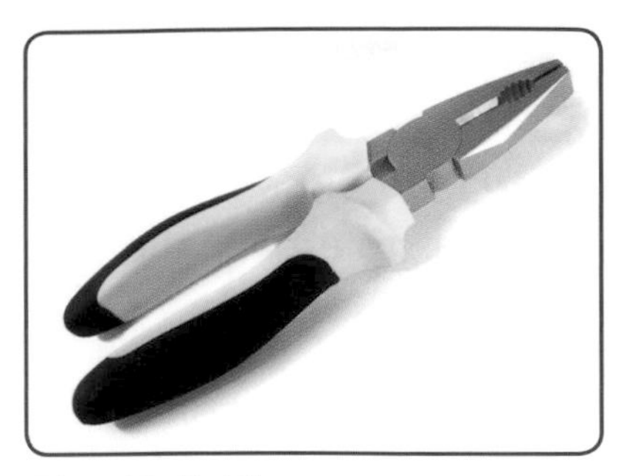

Fig. 7.4 Pliers

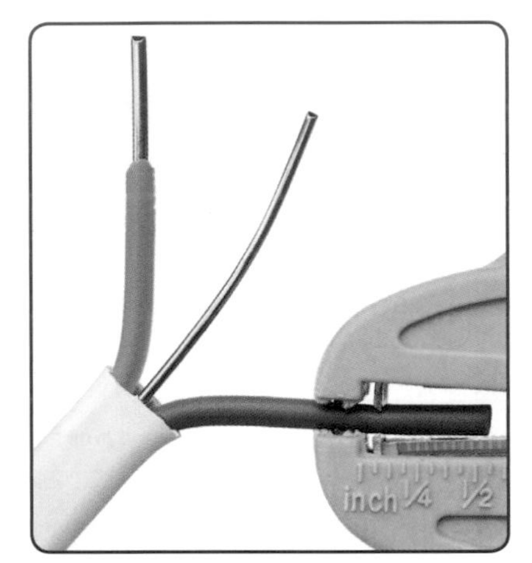

Fig. 7.5 Insulation strippers

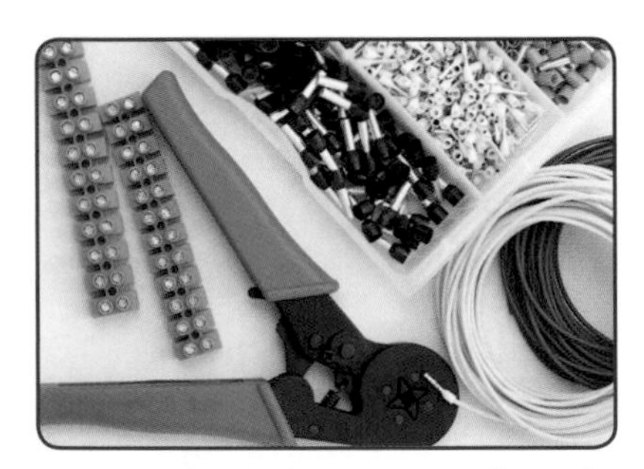

Fig. 7.6 Crimping tool and crimps

together stranded copper wire before making a connection. The curved serrated jaw is designed for gripping round metal rod or tube. It can also be used for holding screwheads or nuts.

- Long nose pliers – these are simply a type of combination pliers specially adapted for working in confined spaces, for example, when placing small washers or nuts on to fittings, and assembling wiring in tight spots.
- Pliers wrench – this is ideal for gripping flat surfaces, such as nuts, without damaging the surface. The jaw can be adjusted to suit the size of the conduit or nuts.
- Insulation strippers – these are used to remove the insulation from around the conductor. Care needs to be taken and the wire strippers set to the correct cutting depth so as not to damage the copper conductor. (Any damage will reduce the current-carrying capacity and size of the conductor.)
- Crimping tools – electrical cables are often terminated using metal lugs that are fitted to the conductor either by soldering or crimping. Part of the lug is squeezed securely on to the conductor using a crimping tool.

TRADE TIP

Never leave tools wet – always dry them and use a recommended lubricant to protect from rust. Never drop tools into your toolbox.

Measuring equipment

As an electrician, you will need to use a range of measuring equipment in order to ensure that cables and support systems are installed level and at the correct distances (i.e. in line with drawings and diagrams). The measuring tools you use will range from simple rulers to laser levelling equipment. However, you need to take care to ensure that all measuring equipment does not get wet or dropped, as this will cause inaccurate readings.

- Steel rulers – used for general measurements of trunking or tray.
- Steel tape measures – used for longer measurements, for example, trunking and conduit lengths or the area of a room.
- Spirit levels – used to check both vertical and horizontal alignment.
- Laser measuring tools – used to measure long distances, for example, one end of a large warehouse to the other end.
- Plumb bobs and chalk lines – used to install and align equipment and accessories vertically. They are weights on the end of a string – when the weight stops moving, the string is at true vertical.

Fig. 7.7 Laser and tape measure

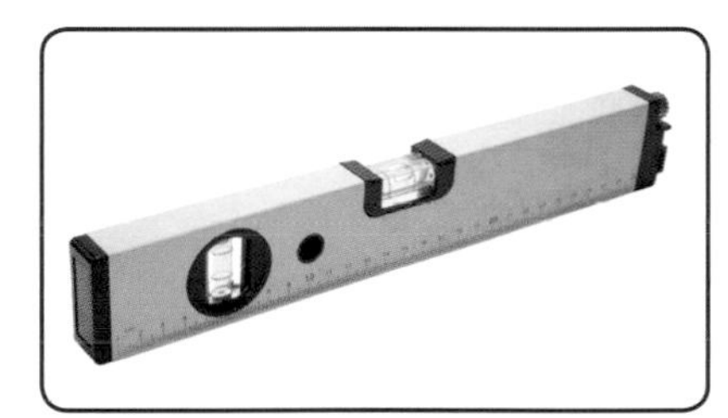

Fig. 7.8 Spirit level

Power tools

Before using any power tools, you will need to carry out the portable appliance testing (PAT) pre-use checks described in Chapter 2, page 49.

- On a construction site, electrical tools usually operate at 110V to reduce the risk of electric shock.

REMEMBER

If electrical equipment is operated at 230V or higher, it should be protected by an RCD (residual current device).

- Portable electric tools should only be used for their designed purpose.
- Never use worn, blunt or damaged bits or other accessories.
- Use battery-operated drills and saws where possible, to prevent electric shock.

Multimeters and test instruments

All test instruments and test leads must comply with the relevant regulations. The HSE has published guidance notes, called GS38, Electrical Test Equipment For Use by Electricians. The BS 7671 IET Wiring Regulations specifies the test voltage and current required to carry out a particular test.

GS38 states that:

- all probes should have finger barriers
- the probe tips should be insulated, to prevent more than 2 mm showing, or spring-loaded retractable screened probes used
- probes should be protected by a 500 mA high breaking capacity (HBC) fuse
- leads should be adequately insulated, of distinguishable colours, flexible and of sufficient length.

Test equipment should be regularly checked and calibrated. Before use, ensure that:

- batteries are in good condition
- there is no damage to the casing

The symbol for alternating current (AC) is:

∿

The symbol for direct current (DC) is:

⎓

- test leads and probes are undamaged and conform to GS38 (see page 182)
- the calibration certificate is valid and current.

Digital multimeters

Digital multimeters are designed to measure alternating current (AC) and direct current (DC).

- AC is generated at 25 kV and stepped down to 230 V for domestic properties using a transformer.
- DC voltage uses battery or photovoltaic (solar) panels.

The meters will measure voltage, current and resistance at different values from millivolts and milliamperes (or milliamps) to kilovolts and kiloamperes (or kiloamps).

- When measuring voltage, the multimeter must be connected in parallel across the load or across the supply, from positive to negative (see Fig. 7.10).
- To measure current in a circuit, the multimeter is connected in series, as shown in Fig. 7.11.
- To measure ohms using a multimeter, you need to set it to Ω and then decide on the value you want – low ohms (mΩ) or high ohms (kΩ).

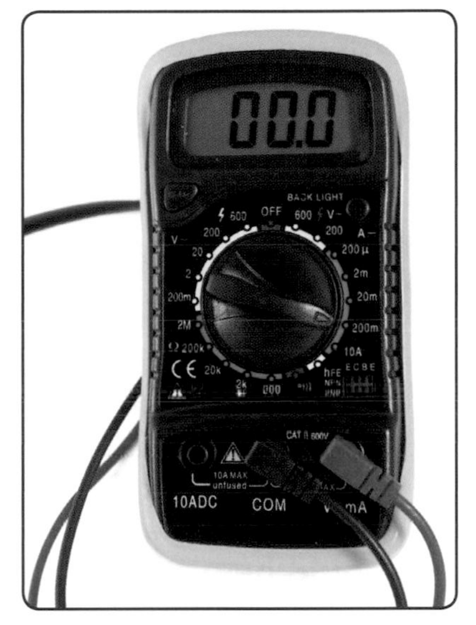

Fig. 7.9 A digital multimeter

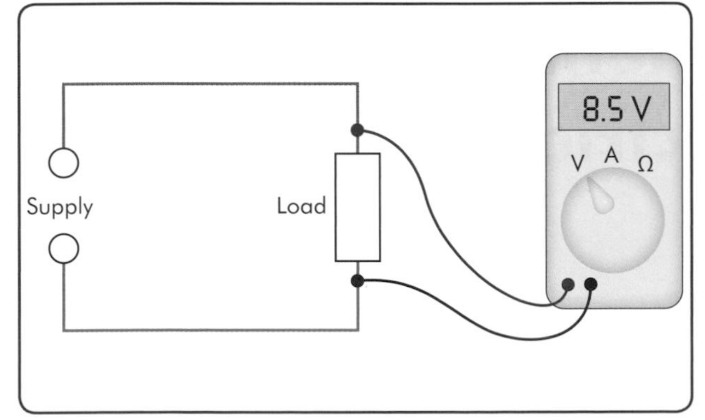

Fig. 7.10 Measuring voltage in parallel

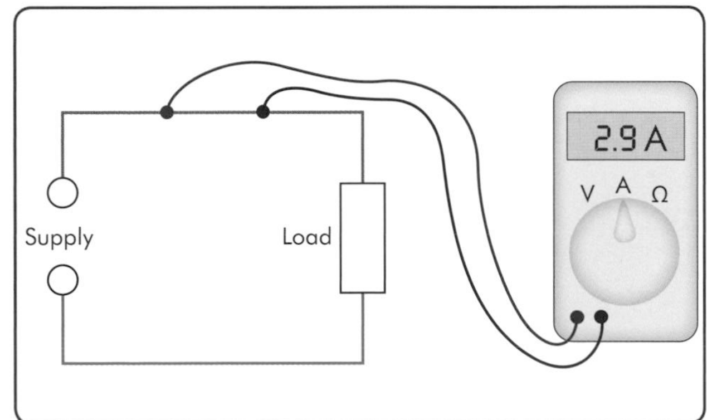

Fig. 7.11 Measuring current in series

REMEMBER

Voltage (*V*) is measured in volts (V).
Current (*I*) is measured in amperes (A).
Resistance (*R*) is measured in ohms (Ω).

TRADE TIP

When terminating cables, it is good practice to double over the end of the conductor. (There should be no exposed copper at the terminals.)

HAVE A GO...

Make up a circuit using a 2.4 V krypton (cycle) light bulb, a lamp holder, a 9 V battery, a switch to control the circuit, and 1 mm^2 conductors or similar.

Then measure the voltage and current.

Next, measure resistance in ohms (Ω) in the circuit at various points. To do this, measure each length of conductor. Check for breaks in the conductors and the resistance across the lamp (this is checking the continuity of the conductors). Remember to attach the positive test probe to the positive terminal and the negative probe to the negative terminal.

You can now calculate your answers to check against your readings using Ohm's law ($V = I \times R$ or $I = V/R$ or $R = V/I$).

If you measure the current and the voltage, you will get: $R = 9\text{V} \div 0.7\text{A} = 12.86\,\Omega$

Materials and components used within the electrical industry

Cables

Cables conduct electricity in circuits and usually consist of a conductor, such as copper or aluminium.

PVC-insulated and sheathed flat wiring cables

This type of cable (manufacturers' reference code 6242) is commonly used for domestic and commercial wiring circuits. It can be sunk into walls and installed in conduit or trunking. It can also be surface-mounted when there is little risk of mechanical damage.

The conductors within PVC-insulated cable have a colour-coded insulator (brown = live and blue = neutral) and a bare copper conductor that is sheathed green and yellow to indicate the circuit protective conductor (earth cable). They are sheathed mechanically, and the sheathing is usually grey or white PVC.

Fig. 7.12 Twin and earth in new colours, including earth sheathing

Single PVC unsheathed cables are used in conduit and trunking installations (see page 188). PVC cable is flexible and the least expensive type. It must not be installed below 0 °C or above 60 °C. The conductors can be solid or stranded.

PVC cables are fixed vertically or horizontally with plastic cable clips, using a cross pein hammer. (The requirements for bends and support can be found in Tables 4A to 4E in the *IET On-Site Guide*).

To terminate PVC/PVC cable you need to follow these steps:

1. Nick the end of the cable with an electrician's knife and pull it apart.
2. When the required length has being stripped, cut off the surplus sheathing with the electrician's knife.
3. To remove the insulation from the conductors, use the wire strippers. Check that the conductors are not damaged and double the ends over with your pliers.

Fig. 7.13 Cable clips

- Stranded conductors should have the strands twisted together to provide a better connection – there should be no bare copper and no loose strands that protrude out of the terminal.

PVC-insulated and sheathed flexible cables

This type of cable falls under manufacturers' reference codes 3092Y and 3093Y.

The construction of the flexible cord is made up of stranded copper conductors insulated in heat-resistant PVC. It is suitable for installations up to 85 °C, so must not be used on heating appliances.

It is important to check that *all* cables conform to British Approvals Service for Cables (BASEC).

General purpose flexible cord

These types of cord (manufacturers' reference codes 3182Y and 3183Y) are available in:

- twin core (brown, blue) used on Class 2 appliances
- three core, where an earth is required (brown, blue, green/yellow)
- four core (black, grey, brown, green/yellow)
- five core (black, grey, brown, blue, green/yellow).

This cable is used to connect domestic appliances to 13 A plug tops and is used for light pendants.

Flexible cable does not require earth sheathing as there are no bare copper conductors.

Wiring systems

There are many different types of wiring systems, enclosures and equipment. They all have to conform to the current BS 7671 IET Wiring Regulations. The regulations require that all electrical installations are designed to provide:

- protection of persons, livestock and property
- the proper functioning of the electrical installation for the intended use.

This type of work is carried out by the designer at the planning stage.

PVC-insulated and sheathed wiring systems

These are used extensively for lighting and power installations in domestic dwellings. To minimise mechanical damage to the cable caused by impact or penetration, the cables:

- must be installed into walls 50 mm below the surface or protected against damage by screws or nails using earthed metal conduit
- must be run within 150 mm of the top of a wall or within 150 mm of the corner and must run horizontally or vertically to the item of equipment.

In addition:

- when installed under floors and above ceilings, cables must be at least 50 mm below the surface or protected against mechanical damage
- when installing into metal back boxes, grommets must be fitted and all sharp edges removed from any accessories used
- cables should not cross over each other or come into contact with gas or water pipes or other non-earthed metal work.

Conduit

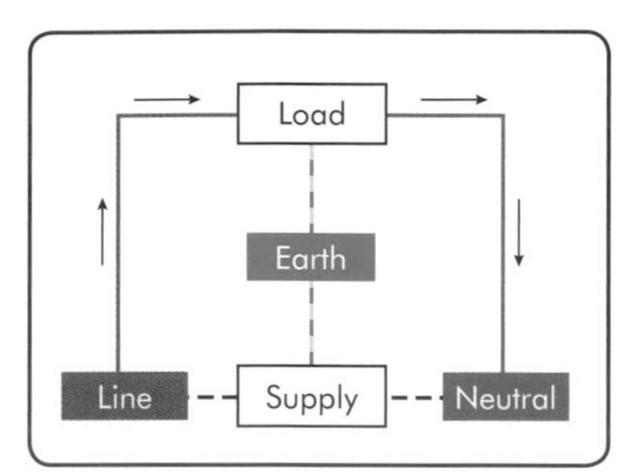

Fig. 7.14 Load supply and earth

Earth or circuit protective conductor

The earth wire is the circuit protective conductor (CPC) and is there to protect you, the equipment and the cable from a fault in the circuit.

You must make sure that the earth connection is continuous all the way through the circuit from the supply to the load.

Steel conduit

The three main types of steel conduit that are commonly used are:

- black enamel conduit, which is used where there is no likelihood of dampness
- stainless steel, which is used in catering environments
- galvanised conduit, which is used outdoors.

PVC non-sheathed cables are run inside the steel conduit, which provides excellent mechanical protection and, in certain conditions, may also provide the means of earth continuity.

The steel conduit is screwed into accessories using special tools called stock and die. These are used to put a thread on to the end of the steel conduit, so that the conduit can be bent to form right-angle bends and sets. The bends are then put into the steel conduit using a bending machine.

Conduit should have a minimum bending radius of 2.5 times the diameter. The amount of PVC non-sheathed cable installed in conduit or trunking should not exceed 45% of the internal cross-sectional area (CSA).

TRADE TIP

When installing conduit or trunking, remove all burrs and sharp edges using a conduit reamer or a file. This will avoid damage to the PVC non-sheathed cables.

Both PVC and steel conduit come in lengths of 3.75 m. Typical diameters are 16 mm, 20 mm, 25 mm and 32 mm.

The types of fittings used are spacer bar saddles, distance saddles, hospital saddles and grampets. A terminal end box is used to attach lamp holders, and through boxes and tee boxes are used to add vertical or horizontal drops to the installation.

Fig. 7.15 Bending machinery and fittings

PVC conduit

There are two types of PVC conduit: flexible and rigid. Expansion couplers need to be fitted to rigid PVC conduit, to allow for expansion of the conduit.

- PVC glue should be used with care to join the accessories together.
- PVC conduit can be cut with a junior hacksaw.
- To put different bends into the PVC conduit, heat is applied to the area and a coiled spring inserted; the bend is then formed using the knee or thigh.

Trunking

Trunking is available in 3 m lengths and a range of CSAs. There are many types of trunking available, for example:

- busbar and rising mains is used for special installations
- special trunking is used for skirting, dado and floor trunking
- PVC mini trunking is often used in domestic dwellings for surface wiring – it can be bought with a self-adhesive backing

TOOLBOX TALK

Fire barriers must be fitted inside the trunking every 5 m, on every floor level or room and any dividing wall.

- steel trunking is used for commercial and industrial purposes – it is fixed using roundhead screws and raw plugs.

The same fittings are used for both PVC and steel trunking. You can buy bends and adaptors for trunking ready formed.

To cut trunking, you need to use a hacksaw; burrs and sharp edges are removed using a file. When joining two lengths of steel trunking together, copper earth straps are used to provide an earth path.

Conduit is often used with trunking as a combined system.

Types of circuits and accessories used with PVC-insulated cables

REMEMBER

All switches and sockets must be installed at a height of between 450 mm and 1200 mm

Within domestic dwellings, there are three common types of lighting circuits and two different power circuit arrangements. These are:

- lighting circuits – one-way; two-way; intermediate
- power circuits – radial; ring main.

Radial circuits

Radial circuits use twin or single sockets. They do not loop back to the consumer unit; they end at the last socket.

- A2 radial circuits are wired in 4 mm^2 PVC/PVC and protected with a 30/32 A MCB, cartridge fuse, RCD or RCBO. The floor area must not exceed 75 m^2.

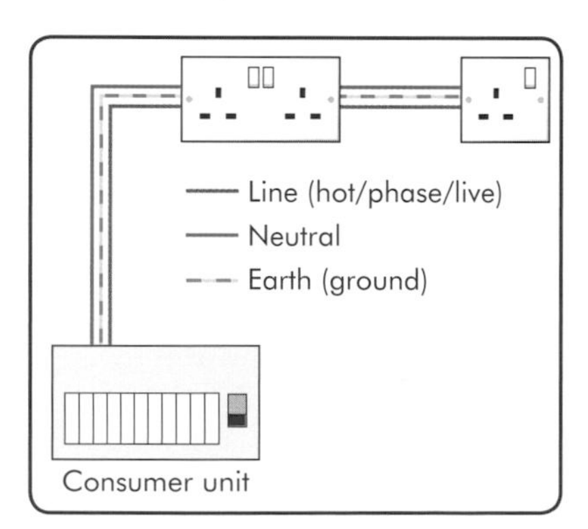

Fig. 7.16 Radial circuit

- A3 radial circuits are wired in 2.5 mm² and protected by a 20 A protective device. The floor area must not exceed 50 m².

Ring main circuits

Ring main circuits start and finish at the consumer unit (line, neutral and CPC). They are connected to each socket in turn, and return from the last socket to the correct terminals in the consumer unit. (Single or twin sockets are classed as one outlet.)

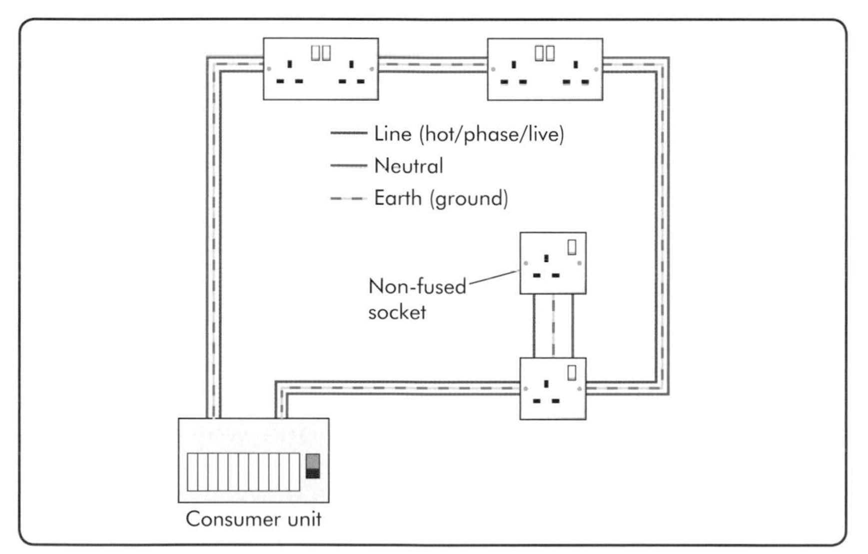

Fig. 7.17 Ring circuit with a spur

REMEMBER

When installing PVC cables, use grommets with metal boxes and remove all sharp edges from PVC back boxes.

Ring main circuits are connected in 2.5 mm² PVC/PVC. A 30/32 A protective device is used. The floor area must not exceed 100 m². An unlimited number of sockets can be installed, provided the regulations are adhered to.

TRADE TIP

Remember to check that all terminals are secure, by gently tugging the conductor, before attaching sockets to the back boxes.

Non-fused spurs

These are limited to the number of sockets or items of fixed equipment on the ring and they must use the same

Fig. 7.18 Double-switched socket

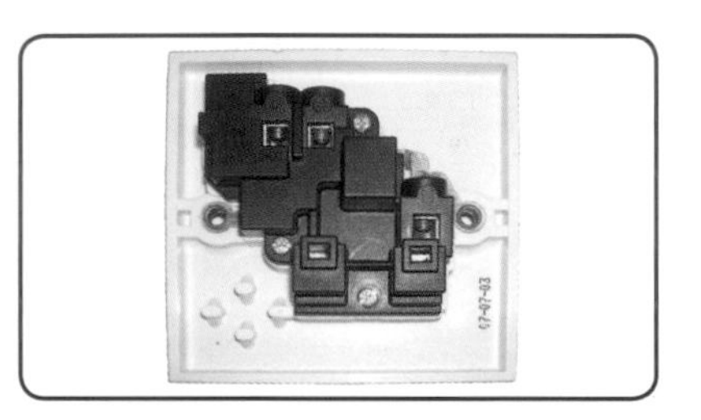

Fig. 7.19 The reverse side of a single-switched socket

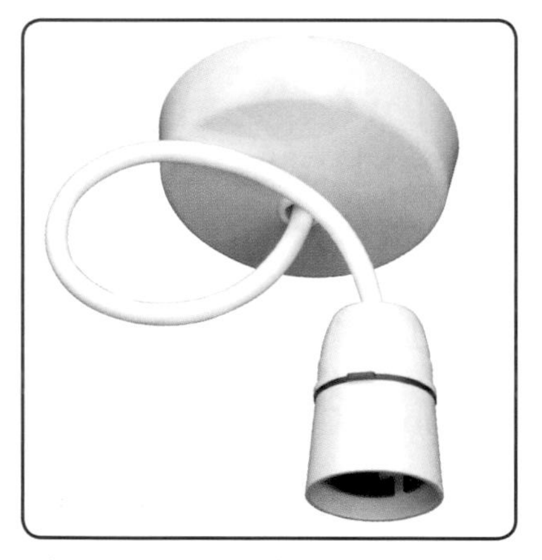

Fig. 7.20 Ceiling pendant lamp holder

Fig. 7.21 Light switch

size cable. So, for example, in Fig. 7.17, you could have three non-fused spurs, single or twin sockets or a fixed item of equipment.

Spurs are connected to the ring directly from the consumer unit by the use of a junction box or from a socket (see Fig. 7.17).

The circuits are constructed using switched socket outlets (single and double) or non-switched sockets, which prevent a circuit being turned off accidently. The sockets can be metal clad or PVC and are attached to metal or plastic back boxes.

Lighting circuits

Lighting circuits are connected using metal or PVC light switch plates and back boxes. They are used for ceiling roses and lamp holders.

One-way circuit

The most basic lighting circuit is the one-way circuit.

A circuit's name depends on the switching arrangement – the one-way switch controls the lamps from one position (see Fig. 7.22). In this case, the switch feed is connected from the fuse to the common (C) in the switch and the switch wire is connected between the L1 switch terminal and the lighting point. The neutral is connected directly to the lighting point.

To connect two lamps, the supply is connected to the switch and a neutral is connected to the lamps. It is a common neutral.

- The switch wire feeds both lamps and they will come on together.
- The two lamps are connected in parallel, so that the supply voltage appears across each lamp.

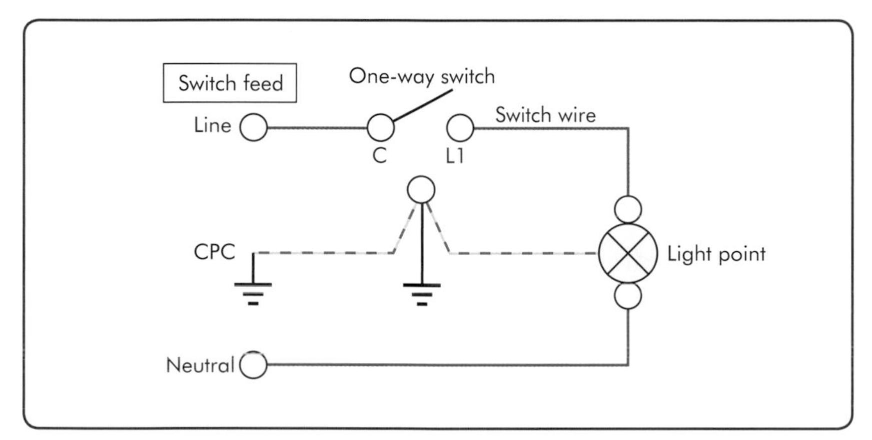

Fig. 7.22 One-way lighting circuit diagram

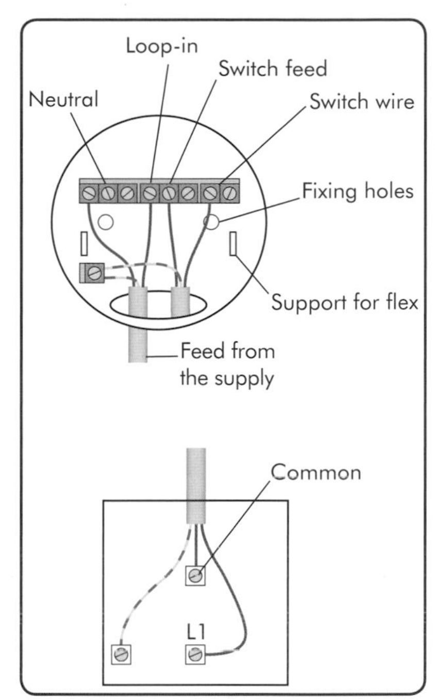

Fig. 7.23 Connections for the loop in method at ceiling rose (not showing the light fitting connections) and one-way switch

Consumer unit

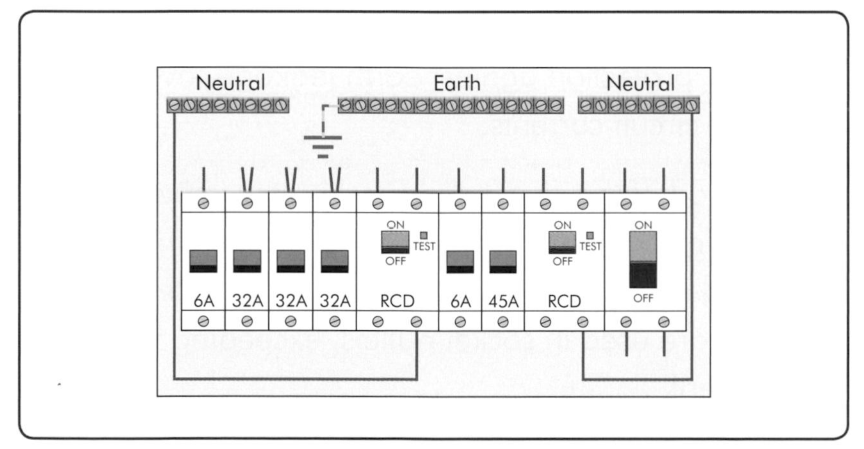

Fig. 7.24 A consumer unit

This should be a 17th edition consumer unit housing 45A, 32A and 6A MCBs to BS EN 60898. It should also include additional protection by means of 80A and 63A/30mA RCD, with a 100A double pole isolator.

REMEMBER

When using metal accessories, ensure all exposed conductive parts are earthed.

The 45 A is used to protect the shower circuit, the three 32 A are for the kitchen, upstairs and downstairs sockets, and the two 6 A are used for upstairs and downstairs lighting circuits.

Protective devices

Residual current device (RCD)

An RCD monitors the current in the line and neutral conductors on a continuous basis. It operates when there is an earth fault; in a healthy circuit, both currents will be equal. When a fault occurs, some current will flow to earth and the line and neutral currents will become unbalanced. The RCD will detect the imbalance and disconnect the circuit.

TOOLBOX TALK

RCD functional testing should be carried out quarterly by pressing the test button.

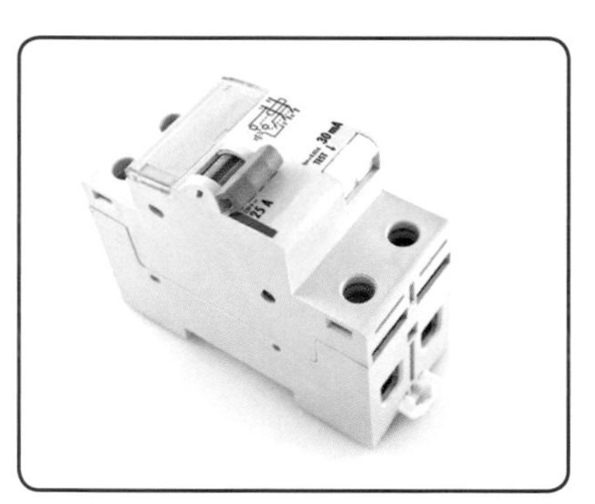

Fig. 7.25 25 A 30 mA RCD

Residual current circuit breaker with overload protection (RCBO)

RCBOs are a combination of an MCB and an RCD. They offer protection against earth leakage, overload and short circuit currents.

An overload of the circuit occurs when too much power is drawn from a healthy circuit, for example when extra load is added to an existing circuit or too many adaptors are used in socket outlets, exceeding the rated load current.

A short circuit occurs when there is a severing of live conductors and insulation breakdown.

Examples of an earth fault include insulation breakdown, incorrect polarity and poor terminations.

Plug tops

13A plug tops use BS EN 1362 fuses to protect the appliance flexible cord. They have three pins. The earth pin is connected to the circuit protective conductor (CPC) of the ring or radial circuit and is longer so that it is the last electrical connection to be removed from the circuit. The other two partially insulated pins are connected to the line (brown cable), which is connected to the fuse, and the neutral (blue cable).

REMEMBER

A fuse protects the cables from damage. BS 7671 Regulation 43-02-01 states that current rating (In) must be no less than the design current (Ib) of the circuit.

HAVE A GO...

Terminating flexible cable into accessories

Using a BS 1363 plug top, terminate a three-core flexible cord.

You will need to:

1. use the stripping knife and *lightly* score around the flex's outer sheathing
2. bend the flex where you have scored it (this should part the outer sheathing)
3. pull the outer sheathing off (you should not be able to see any copper where you have used the knife)
4. use the wire strippers to strip back the ends of the wire around 10 mm to reveal the copper (make sure the strippers are set to the correct cutting depth)
5. twist the stranded copper conductors tightly and insert into the correct terminals (brown = line; blue = neutral; green/yellow = CPC).

Note: the earth wire in a flexible cable already has sheathing fitted.

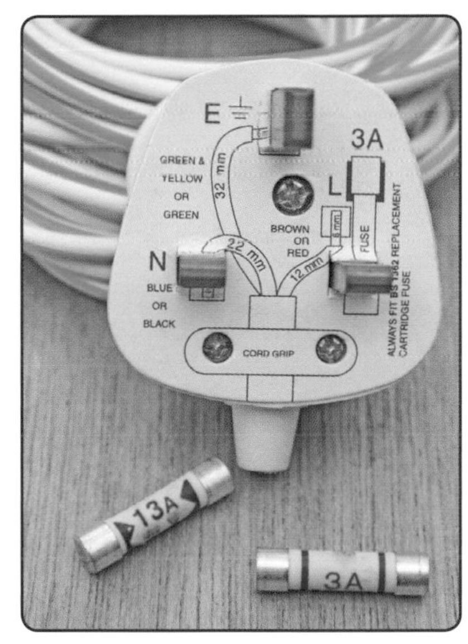

Fig. 7.26 13A plug top with conductor lengths

REMEMBER

Always use the cable grips in plug tops and ceiling roses.

HAVE A GO...

Select an electrical piece of equipment and look at its power rating. Calculate the fuse required.

Working out the correct fuse required

It is important to check that you are using the correct BS 1362 fuse. This should be either a 3A red up to 700W, a 5A black up to 1000W (1kW) or a 13A brown up to 3000W (3kW).

Power (P) in a circuit is found using the formula $P = I \times V$. As you want to find current (I), you need to transpose the formula to $I = P/V$.

For example, for a 1300W drill supplied by 110V: $I = 1300\text{W}/110\text{V} = 11.82\text{A}$

You would therefore use a BS 1362 13A cartridge fuse.

Basic electrical practical applications

Common types of fixings

Within the electrical industry, a large variety of fixings are used. These range from girder clamps used on RSJs to raw bolts used to secure equipment to walls and floors.

The various types of fixings and their uses include:

- wood screws and plasterboard fixings, fast fix, spring toggles and gravity toggles are used on cavity walls
- masonry nails are used with PVC clips
- galvanised clout nails are used for fixing capping onto masonry walls
- pop rivets are used to install two pieces of metal together

- different types of threaded metal bar and wire cables are used to support or suspend cable systems (basket tray or lighting trunking)
- wood screws are used to secure equipment to wood or, when used with plastic raw plugs, to fix equipment to masonry.

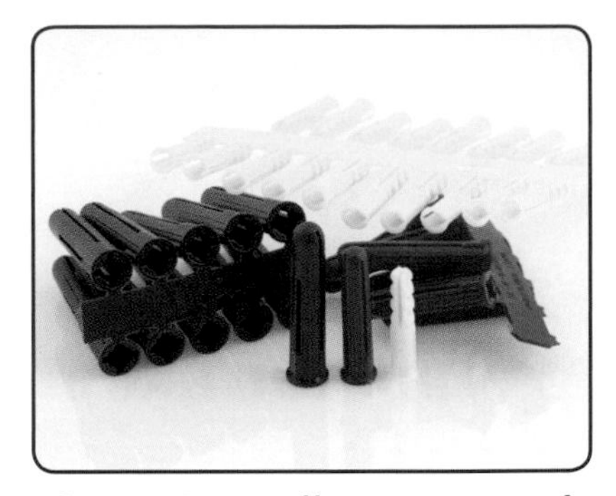

Fig. 7.27 Different types of raw plugs

Different types of raw plugs have different types of heads. Those with round heads are used for trunking, while those with counter sunk heads are used for spacer bar saddles or when accessories have counter sunk recess.

Screws are selected by the shank diameter, length, type of head (cross, slotted) and the metal finish used for the type of environment, for example, non-ferrous is coated with rust prevention.

Cable clips are used to secure PVC cable to walls. It is good practice to place the nail part underneath the cable when installing horizontally, to give additional support.

Diagrams and drawings

There are many different types of electrical drawings and diagrams, for example, circuit, schematic wiring, layout, and block diagrams.

Block diagrams

This is the simplest type of diagram. Squares or rectangles are used to represent pieces of equipment.

Block diagrams show how the components of the circuit relate to each other, but do not show details of termination.

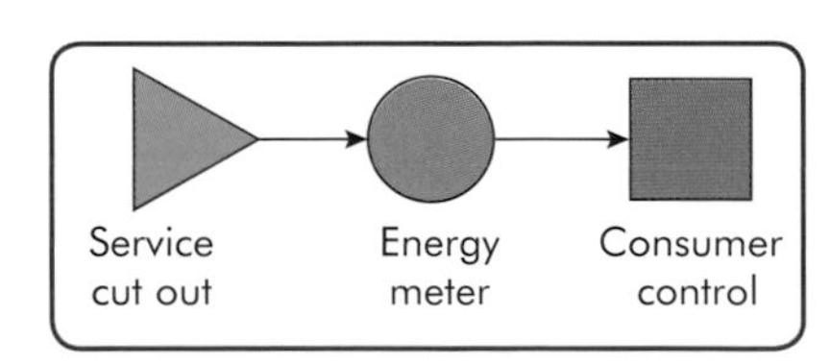

Fig. 7.28 Block diagram

Circuit diagrams

Circuit diagrams are very similar to schematic diagrams and use BS EN 60617 symbols. They show how the circuit is connected but not the physical connections.

DID YOU KNOW

A common ratio for drawings is 1:50. So 10 mm on a drawing using this ratio would represent 500 mm in real life.

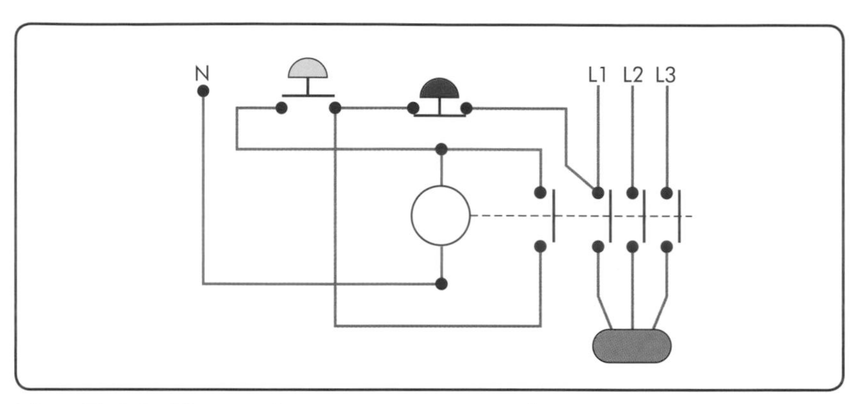

Fig. 7.29 Three-phase motor circuit diagram

Schematic wiring diagrams

These show how the circuit is wired. They include a physical layout of the cable route and the accessories.

HAVE A GO...

1. Produce a circuit and wiring diagram for a two-way one-light circuit.
2. Using BS EN symbols, produce a layout diagram for a room in your house.

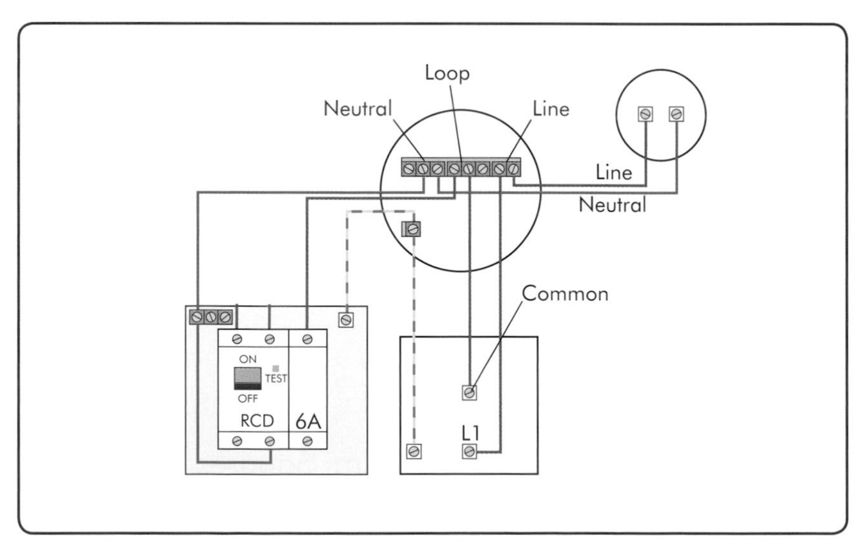

Fig. 7.30 One-way wiring diagram. (This shows a simplified consumer unit for clarity)

Layout diagrams

Layout diagrams use BS EN 60617 symbols and are scaled drawings (common scales are 1 : 50 and 1 : 100).

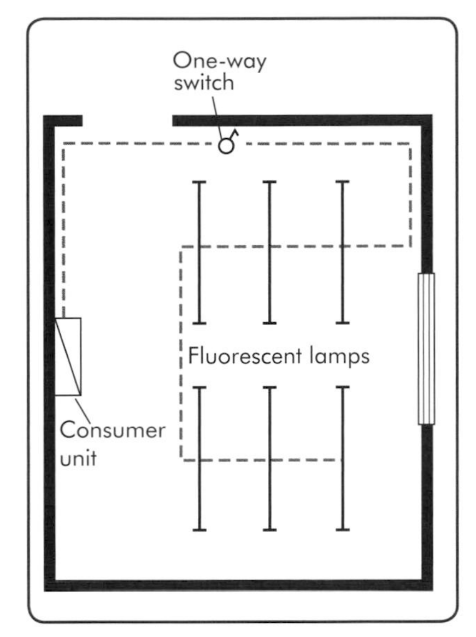

Fig. 7.31 Layout diagram with six fluorescent lamps, a consumer unit and a one-way switch

Safe isolation procedure

Before testing a circuit, you need to ensure that the circuit is dead and isolated from the supply.

1. Locate and identify the circuit or equipment to be isolated and the means of isolation (this could be the fuse or isolation switches).
2. Lock off the electrical supply and place a warning sign (e.g. 'Safety electrician at work').
3. Select an approved voltage indicator or test lamp.
4. Check that the device is functioning correctly on a known supply or a proving unit.
5. Check the circuit or equipment to be worked on is dead using the approved voltage indicator or test lamp. (Test line to earth, line to neutral, neutral to earth.)
6. Recheck the approved voltage indicator or test lamp on a known supply or proving unit.

Fig. 7.32 Kit for safe isolation

REMEMBER

SOLO (Switch Off, Lock Off).

Always carry out safe isolation before starting work.

Continuity of the CPCs test

Two tests are used to check that the conductors are electrically sound and correctly connected.

Link method

1. Link the line conductor and the protective conductor together at the consumer unit.

2. Use the low ohmmeter to test between the line and earth terminals at all light switches, light fittings and equipment.
3. Record the highest value of R_1 and R_2 on the installation schedule.
4. Use the value of R_1 and R_2 to calculate the earth fault loop impedance.

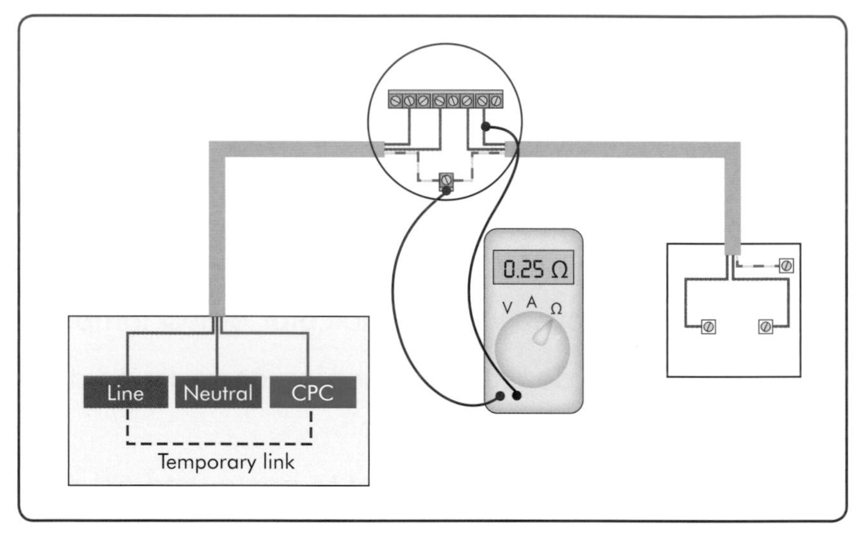

Fig. 7.33 Link method of testing

Long lead method

This method is used to obtain continuity of the bonding conductors.

1. Connect one lead to the consumers' main earth terminals, and the other lead to a trailing lead.
2. Test the protective conductors at light switches and light fittings.
3. Record the highest value of R_2 on the installation schedule.

When using both test methods to check continuity of the CPC, it is important to remove supplementary bonding to prevent false readings from parallel paths.

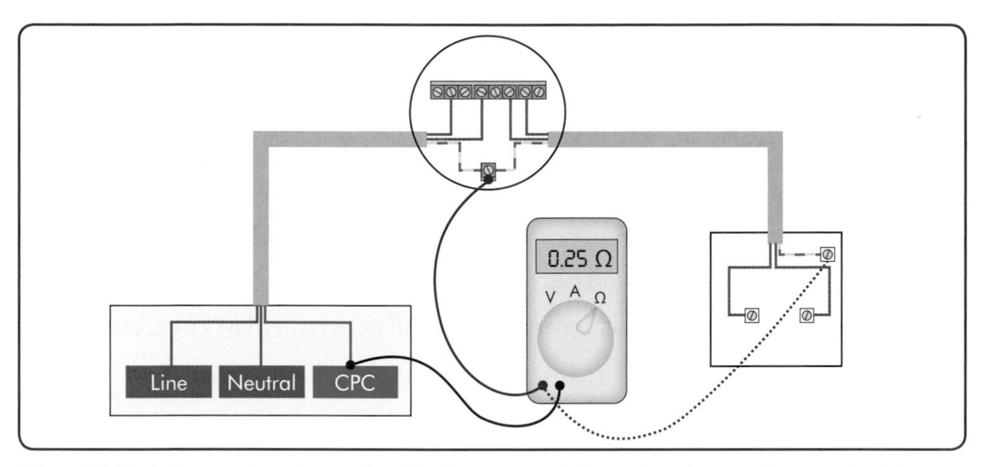

Fig. 7.34 Long lead method of testing (also check earth in switch)

Continuity of ring final circuit conductors test

The continuity of each conductor, including the CPC of every ring final circuit, must be checked to ensure that the ring is complete, has no interconnections and is not broken. The test is carried out in three steps.

1. Test the line, neutral and CPCs end to end at the consumer unit. Assume R_1 is the line, R_2 is the earth, and R_n is the neutral. If wired in PVC/PVC twin and earth, the reading for the earth conductor should be 1.67 times greater than R_1 (within 0.05 Ω), when the same size conductors are used.

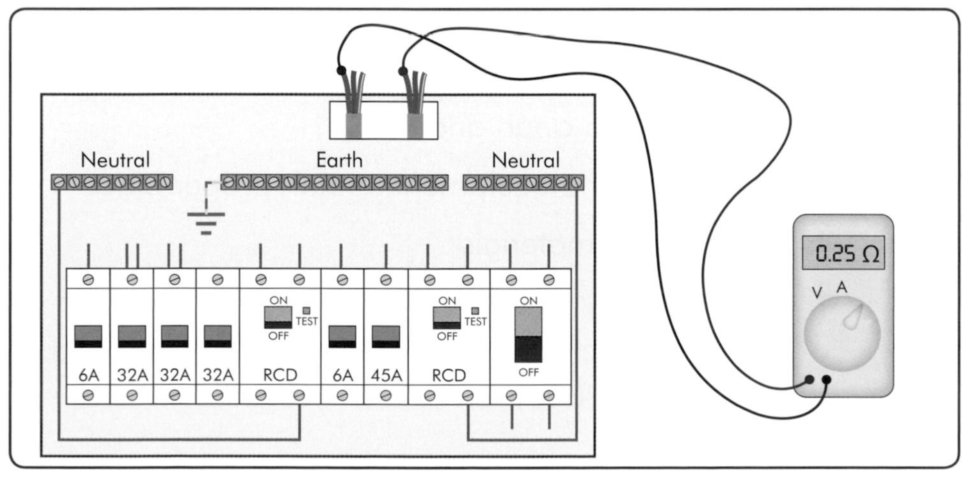

Fig. 7.35 Continuity of ring testing step 1

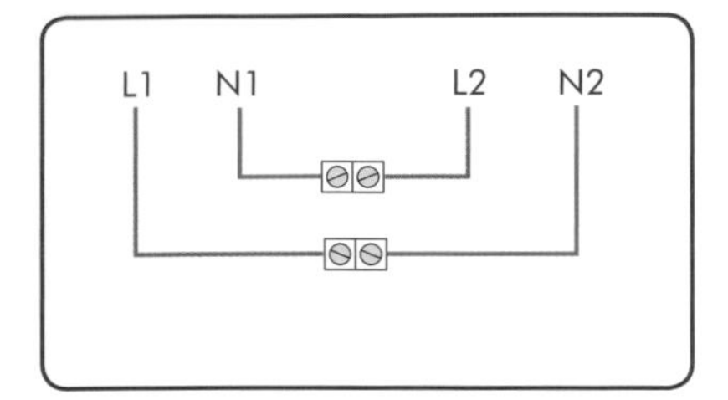

Fig. 7.36 Continuity of ring testing step 2

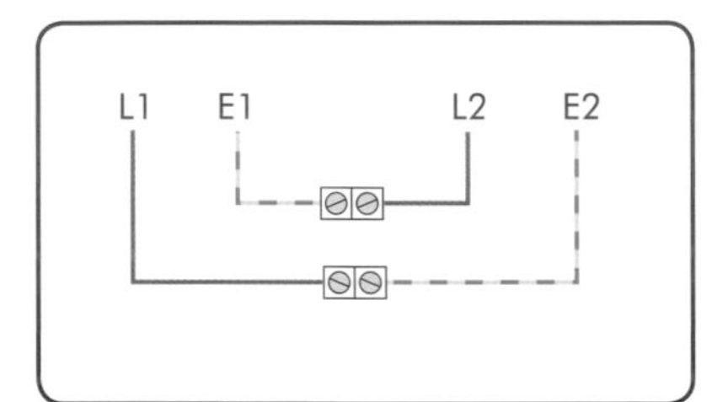

Fig. 7.37 Continuity of ring testing step 3

2. Connect the line and neutral conductors together so that the outgoing line (L1) is connected with the returning neutral (N2) and the outgoing neutral (N1) is connected to the returning line (L2). Measure the resistance of the line and neutral conductors at each socket outlet. The readings obtained should be mostly the same; the measured value would be a quarter of (R_1 + R_n) if readings are obtained in step 1.
3. Repeat the previous steps with the line and CPC cross connected at the consumer unit and test each socket. Record the highest resistance value of R_1 + R_2 on the installation schedule. This test also confirms polarity of each socket outlet and can be used to determine the earth loop impedance value of the circuit. The measured value at each socket outlet and at the consumer should be mostly the same; the measurement would be a quarter of (R_1 + R_2) if readings are obtained in step 1.

Completing the task

It is important that you ensure that the work area is left in a safe condition after the task. You will need to make sure that you:

- leave the area clean and tidy
- return tools and equipment to their proper place
- return excess materials
- dispose of any waste materials in accordance with the Waste Electrical and Electronic Equipment (WEEE) Regulations (2006).

CHECK YOUR KNOWLEDGE

LEVEL 1

1. What tool would you use to remove insulation from a conductor?
2. Name three checks to be carried out on test equipment before use.
3. What is measured in parallel and what is measured in series?
4. What is measured in ohms? Draw the ohms symbol.
5. At what height should sockets and switches be installed in domestic properties?
6. A ring main wired in 2.5 mm^2 and 1.5 mm^2 PVC/PVC should have what size of protective device?
7. Explain why lighting circuits are connected in parallel.
8. Explain why the earth pin of a 13 A plug top is longer than the line and neutral.

8 Heating and ventilation operations

This chapter covers the learning outcomes for:

Understand and demonstrate fundamental heating and ventilating operations

City & Guilds unit number 107; EAL unit code QACC1/07; ABC unit code A06 for Level 1 and A02 for Level 2.

This chapter is about the basic requirements and principles of heating and ventilation systems, whether in new or existing buildings.

When designed and installed correctly, heating and ventilation systems are able to provide a comfortable environment economically. There is a wide range of systems that provide heating and domestic hot water, and they depend on a number of factors, such as type of building, use and occupancy, and the fuel to be used.

However well designed a system may be, if the system is to perform as intended it is crucial that its installation matches the design and specification. To this end, tests and commissioning checks will highlight any defects in the way the installation has been set up or designed.

IN THIS CHAPTER YOU WILL LEARN ABOUT:

- regulations applicable to heating and ventilation systems
- health and safety for heating and ventilation systems installation
- basic hand tools used to carry out heating and ventilating work
- heating and ventilating materials and components
- basic installation of low carbon steel pipwork
- commissioning and testing systems.

Health and safety for heating and ventilation systems installation

Heating and ventilation systems vary and consideration must be given to all involved in their installation. In addition, while work is ongoing there may be other trade operatives working on the same site, as well as the client and any visitors. For these reasons, health and safety must be a primary consideration.

These are some key health and safety points when installing heating and electrical systems:

- As heavy components need to be transported and lifted into position, foot protection is essential.
- Drilling and fixing at various heights and angles requires eye protection.
- Many traditionally heavy manual aspects have been improved by the introduction of electrical tools and equipment to make work easier, faster and safer to carry out. Therefore, you need to have a sound understanding of the safe use of electrical tools and equipment.
- Prevention of accidents is essential, so you will need to know how to carry out risk assessments to ensure that sufficient controls are in place. With heating and ventilation work, each job varies and it is important to allow for this when undertaking risk assessments.
- You will also need to be familiar with health and safety legislation, such as the COSHH Regulations.

DID YOU KNOW

To work on building sites, operatives need a current Construction Skills Certification Scheme CSCS card. To find out more, go to: www.cscs.uk.com/health-and-safety-test

TRADE TIP

Work on new buildings is often referred to as 'site work', while work on existing buildings is generally referred to as 'refurbishment work'.

TOOLBOX TALK

Before you can use some equipment, specific training and certification may be required.

ACTIVITY

List five items of basic personal protective equipment (PPE) that an operative would require for working on site.

REMEMBER

Look back at Chapter 2 to remind yourself of how to carry out risk assessments, key health and safety legislation, and using method statements.

- Another useful tool to reduce the risk of accidents and improve efficiency is a method statement. This is an agreed and documented list of how a task or job should be carried out and the methods to be used.

QUICK QUIZ

When working on a building site, what would you need to take with you every day?

a) Hard hat and high-visibility vest
b) Hard hat, high-visibility vest and boots
c) All PPE equipment required for the work being carried out and your CSCS card
d) PPE and a form of identification

Basic systems

The systems installed range from air conditioning and heating systems using traditional boilers to the latest technology using air or ground source heat pumps. The newer systems provide heating and/or cooling by extracting heat from the ground or outside air and then distributing the heat around the building using either wet systems, such as underfloor heating or radiators, or dry systems using ductwork.

KEY TERMS

HVAC: The abbreviation for heating, ventilation and air conditioning.

Duct work

Heating and ventilation engineers may be involved in the installation of duct work. These systems are used on large **HVAC** systems to transfer air for heating or cooling of the internal environment. They may also involve the installation of various hot and cold water systems for the building and to ensure energy efficiency. Duct work may also be used to move ventilation air to and from buildings.

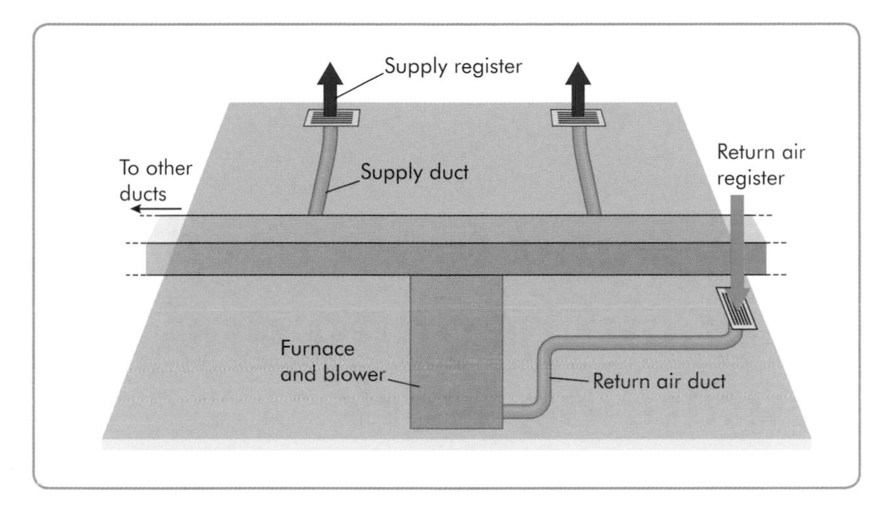

Fig. 8.1 A simple ductwork system

Chilled water

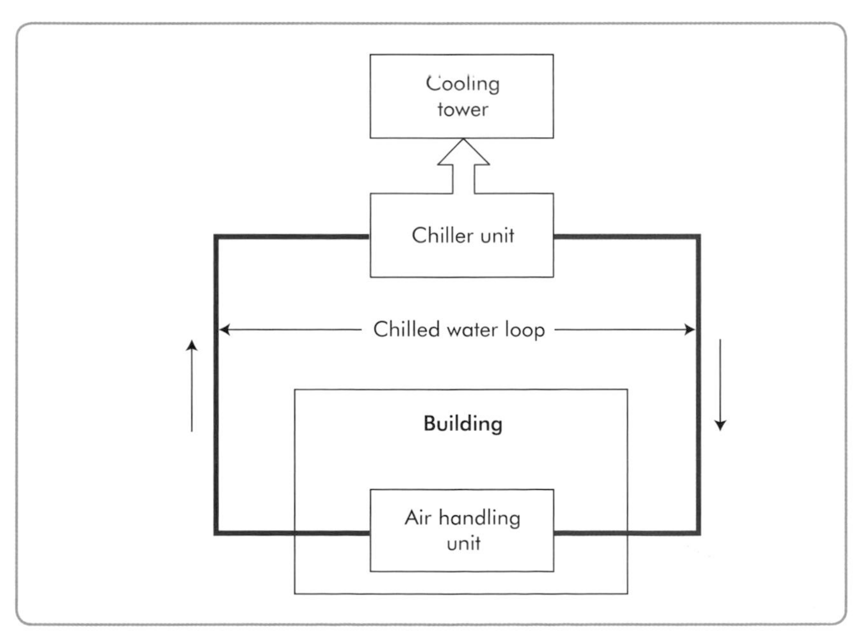

Fig. 8.2 A simple chilled water system

DID YOU KNOW

Fuel costs are increasing rapidly for traditional fossil fuels, making it even more important that heating systems are run efficiently. To find out more, go to: www.carboncounted.co.uk/when-will-fossil-fuels-run-out.html

HAVE A GO...

Find out which fossil fuels are used for heating and ventilating systems.

TRADE TIP

Many 'chillers' are used for air conditioning applications and are too large and heavy for domestic situations.

Chilled water systems are essentially cooling systems. They circulate chilled water around the building or through cooling coils. The system removes heat from the water, which is then circulated through a heat exchanger to cool a building's air (or sometimes equipment). Chilled air is then recirculated back to the chiller to be cooled again.

Basic hand tools used to carry out heating and ventilating work

TRADE TIP

You are generally expected to work within tolerances of plus or minus 3 mm when installing pipework.

A wide range of general tools, such as hammers, chisels and drills, are used for heating and ventilation work. Other tools are more specific to heating and ventilation, for example pipe cutters.

Basic tools

Metric tape measures

These are available in lengths from 3 m to 6 m. A tape measure is a crucial tool for installation work and must be maintained in good condition so it can be easily used and read.

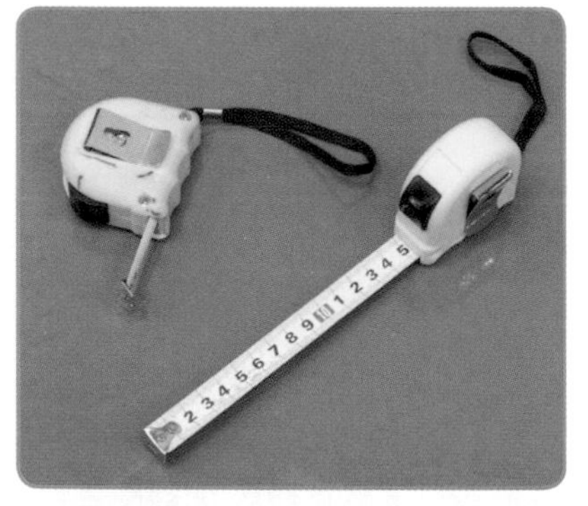

Fig. 8.3 Tape measures

Spirit levels

These come in various lengths, from 150 mm (boat level) to 1000 mm (available up to 2 m). They are used for setting out and marking positions for pipework clips (some need to be perfectly level or with a rise or fall depending on the

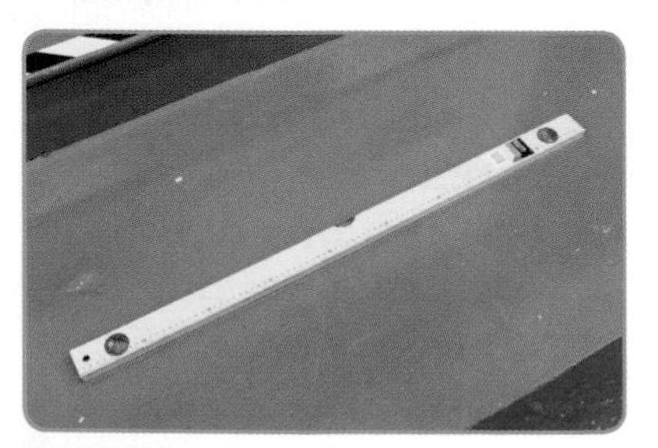

Fig. 8.4 Spirit level

system) and for marking out for fixings for components such as radiators and boilers. Line levels hung onto a string line for taking levels over longer distances are available, as are laser levels, which are useful for marking out levels over long distances. Spirit levels must be kept clean and their accuracy checked regularly.

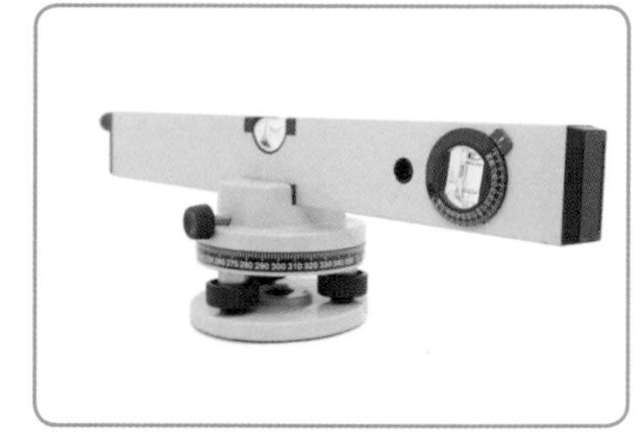

Fig. 8.5 Laser spirit level

Screwdrivers (cross and flat head)

Screwdrivers should be kept in good condition and the appropriate size used, so that the tip of the driver fits snugly into the head of the screw.

The handle of the screwdriver should be secure and both the handle and tip should be kept in good condition. If not, the screwdriver should be replaced.

When working on electrical equipment, only insulated and tested electrical screwdrivers should be used.

Fig. 8.6 Different types of screw heads

TRADE TIP

Many operatives use battery-powered interchangeable-tip screwdrivers to increase productivity and reduce the number of tools to be carried.

Hack saws

These are used for cutting pipes and other metal materials. Two types of hacksaw are used: the larger frame type and the smaller junior hacksaw, which is often used for cutting into existing pipework, copper and smaller diameter tubes.

TOOLBOX TALK

When using a hacksaw, check that the material to be cut is secure, that your posture is correct, and that your method produces square cuts.

Various blades are available for cutting different materials. It is important that the right blade is used and tensioned correctly within the frame, with the blade teeth facing forwards (away from the handle). The cut is then made on the forward stroke of the cutting action. You will also need to ensure that the cut is made square and any swarf removed from inside the tube.

KEY TERMS

LCS: Low carbon steel tube. It is often referred to as black iron pipe or mild steel pipe or, if zinc coated, as galvanised iron pipe.

Blades having:

- 32 teeth per 25 mm of blade are used to cut low carbon steel (**LCS**) pipe
- 22 teeth per 25 mm of blade are used to cut copper tube.

Pipe cutters

These tools come in various sizes and designs, with some specifically designed to allow installed pipework to be cut in its existing position.

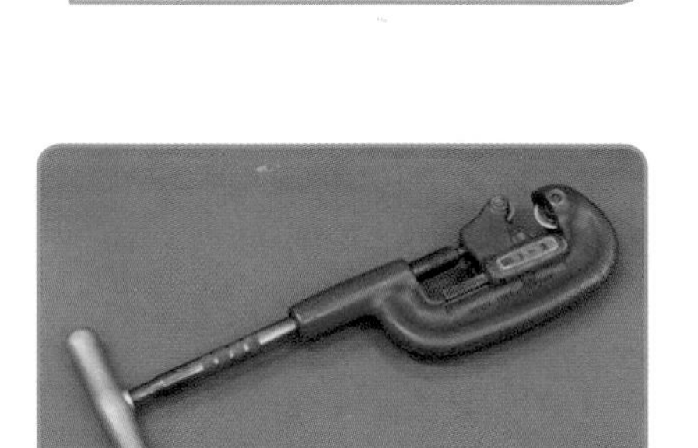

Fig. 8.7 Pipe cutters for LCS tube

Pipe cutters are often used instead of hacksaws as they guarantee a square cut each time if they are maintained and used correctly. They also remove much of the hard work and effort required in using a hacksaw when cutting large diameter pipes.

TRADE TIP

Too much pressure on the cutting edge of the pipe cutter blade increases the risk of the blade spiralling along the tube rather than cutting through it.

The main disadvantage of using pipe cutters is they reduce the internal pipe diameter as they cut through the tube. This reduction has to be removed to bring the internal diameter back to its original size.

When using a pipe cutter, you need to take care to avoid excess pressure during cutting, as this blunts the blade and significantly reduces the internal bore.

It is important to inspect and maintain pipe cutters regularly. This will ensure that the replaceable blade or blades remain sharp, and that the pipe rollers and adjustable jaws are lubricated and can move freely.

Pipe reamers

These are available in various forms, but the reamers we are interested in are for use on LCS tube. Of these, some are designed to be used on one diameter of tube only, although most are designed to be used on 10–32 mm tube diameters.

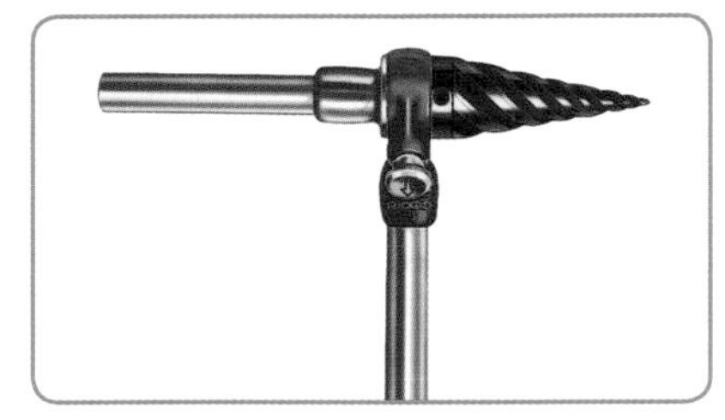

Fig. 8.8 Reamer for LCS tube

The most commonly used type of pipe reamer is a ratchet reamer. These save time and effort when removing burrs from the inside of a tube after it has been cut with pipe cutters. Their use helps to ensure that:

- the internal pipe diameter is the intended size
- there is no disruption to the flow of the liquids through the tube
- the chance of any debris in the system causing blockages is reduced.

To maintain pipe reamers you need to:

- sharpen the cutting edges of the blades
- check the ratchet mechanism (accidents can occur if the mechanism slips while in use)
- check that the handle is in good condition and secure.

Files

Files are made from cast steel and come in many shapes and sizes, from 100 mm to 350 mm. They are

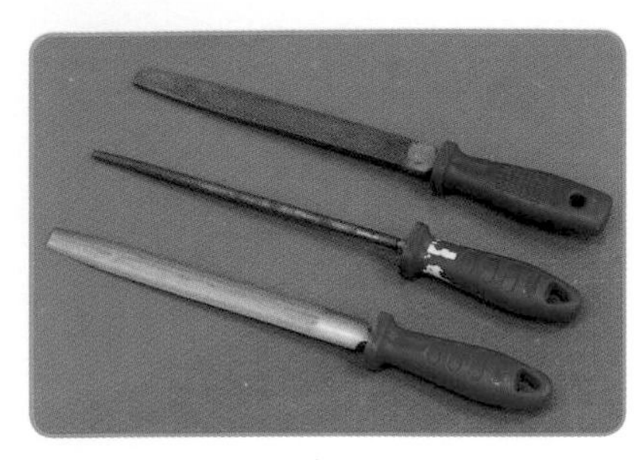

Fig. 8.9 Half-round, round and flat files

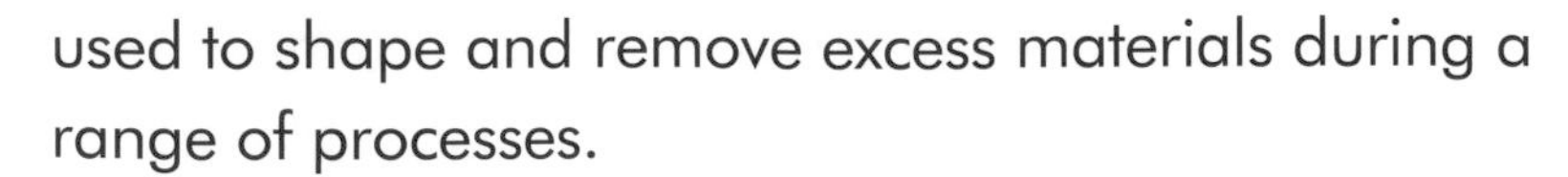

used to shape and remove excess materials during a range of processes.

- Half-round or round (rat tail) files can be used to remove burrs from inside a tube after cutting.
- Flat files are used to remove external rough edges after cutting with hacksaws.
- The face of a file is usually a double cutting surface, while edges and round surfaces are single cut.

To maintain files you need to:

- check the condition of the cutting surfaces (clean with card wire)
- check the condition of the handle
- check the fixing of the tang (handle end of the file shaft) into the handle.

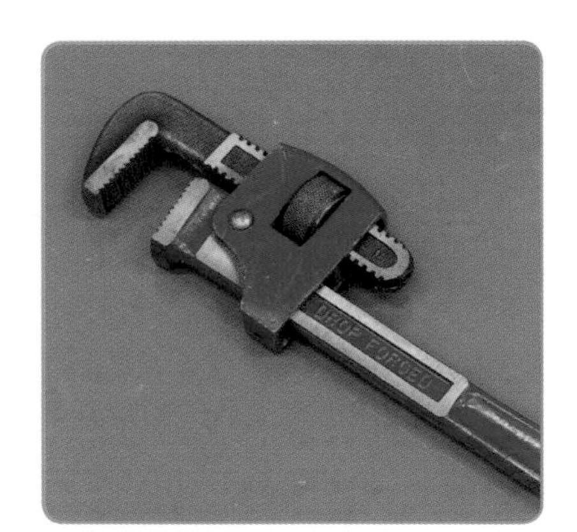

Fig. 8.10 Heavy-duty pipe wrench

Pipe wrenches

Pipe wrenches are the most widely used tool for assembling LCS pipework. They are available in sizes that refer to their length (150 mm to 600 mm, though larger sizes are available for large diameter industrial pipework). They are traditionally made of steel and are heavy to use. However, today they are available in aluminium alloys, making them considerably lighter.

The standard jaws are designed to give a good grip on LCS tube when assembling threaded joints, and the most commonly used size of 450 mm will cope with tube diameters up to 50 mm (2″). If used on unsuitable materials, such as chromed tube, they will cause serious marking or damage to the tube or fitting, so an alternative tool should be used.

It is important to correctly adjust and check jaws before applying pressure, as this will avoid any danger of slipping and injury.

Pipe wrenches are available with flat-faced jaws. Angled heads are specifically designed for use on flat-faced surfaces found on some fittings and valves.

Fig. 8.11 Chain wrench

TRADE TIP

Pipework may be difficult to assemble in position with a pipe wrench. It may be better to use a chain wrench, as it needs less space.

Adjustable spanners

Various sizes and shapes of spanners are available for use on flat-face surfaces, such as nuts found on fittings and valves. It is important to ensure correct adjustment of these tools, to prevent slippage during use and damage to the surface of the nuts or surrounding materials.

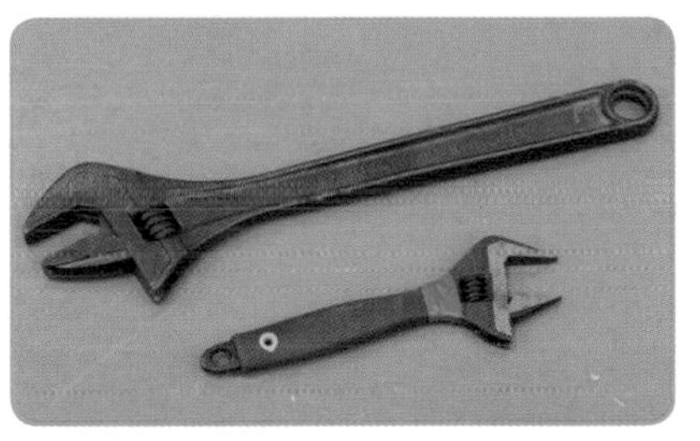

Fig. 8.12 Adjustable spanners

Stocks and dies

These are used for cutting threads on the end of LCS tube, for jointing with malleable iron fittings.

There are several types of manual stocks and dies available, which are capable of cutting either tapered or parallel threads (see Fig. 8.15). Fixed, interchangeable (drop head) or adjustable die heads (chaser die stocks) can cut a range of thread sizes but require adjustment or change of cutting dies for different pipe diameters. Most **stocks** incorporate a ratchet system to turn the die head.

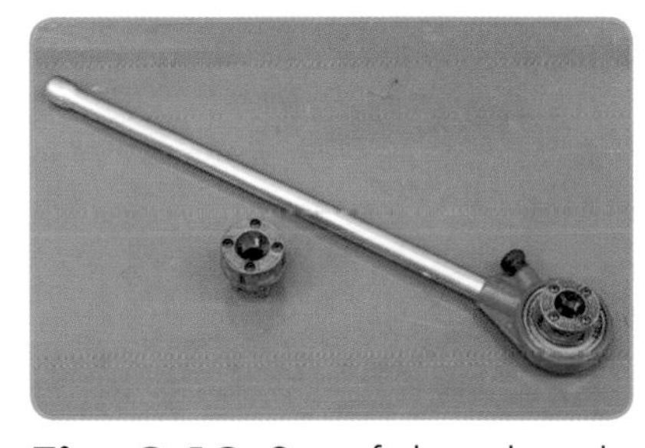

Fig. 8.13 Set of drop head dies

KEY TERMS

Stocks: The die-holding section of the tool.

BSP: British standard pipe

Fig. 8.14 Chaser dies stocks showing the method of adjustment

BSP tapered threads are commonly used on LCS tube used for jointing. The most widely utilised stocks and dies for jointing by hand are drop head dies; these are available in kits for cutting threads from 6 mm (⅛″) up to 50 mm (2″). The cutting dies are permanently fixed into the die head and need no adjustment. Die heads for different size pipes can be quickly changed.

Electric hand-held dies are also available and allow threads to be cut on pipes in position.

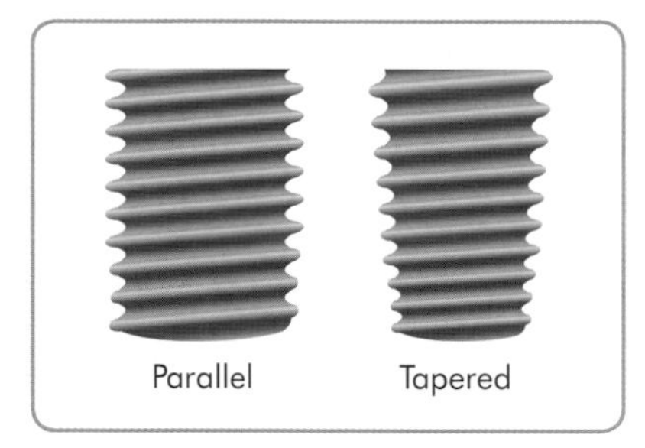

Fig. 8.15 Tapered thread and parallel thread

Fig. 8.16 Compact electric threading machine

DID YOU KNOW

Thread sizes are still referred to using imperial sizes (1″ is equivalent to 25 mm).

ACTIVITY

1. List the advantages of using drop head dies over any other type that has to be adjusted when changing tube size to be threaded.
2. List four health and safety considerations if you were working on a building site and had to cut a thread on a long length of LCS pipe being held in a stand-type pipe vice.

TRADE TIP

Electric threading machines are normally used for large contracts on site. They allow the tube to be cut, reamed and threaded in one three-stage operation.

Hydraulic bending machines

These are used to make bends and offsets in LCS tube. Several types are available, but the most widely used are those shown in Fig. 8.17. These can be stand-mounted, used on a floor or bench-mounted. Although they are very heavy when assembled, they are designed to be transported easily once disassembled.

1. Measure and mark out where the tube is to be bent.
2. Select the correct size former (different formers are used for LSC and copper).
3. Place the tube in the former located at the end of the hydraulic ram shaft.
4. Fit the tube corner stop and pins in their correct position (the tube could be damaged if they are placed incorrectly).
5. Check that the vent/breather valve or control is in the 'open' position.
6. Pump the hydraulic ram handle up and down to advance the former and tube.

TRADE TIP

When using hydraulic bending machines for 90° bends, the angle needs to be over bent by approximately 5°, as the pipe will spring back by this amount.

TOOLBOX TALK

Never put your hands, fingers or body parts in a position where they could be crushed during the bending operation.

Fig. 8.17 Tip wing hydraulic bending machine for LCS tube

TOOLBOX TALK

Never hold LCS tube in a flat-face jawed vice to assemble joints. The vice will not hold against the pressure required to tighten the joint and could cause an accident or injury.

7. Hold the tube against the corner pipe stops as the ram advances and bends the tube.
8. Pump again, until the desired angle is bent.
9. When the desired angle has been bent, pump the handle once more.
10. Release the pressure on the ram and pump handle to withdraw the former from the tube stops.
11. Remove the tube from the former. The tube may become wedged in the former while bending, but you must never strike the former with a hammer to remove it. Instead, place a wooden block on the floor and strike the end of the tube onto the block while holding the former, which will be released.

You can straighten over-bent angles by up to 10° by using a flat-faced straightening former with the tube bend held in place by the pin corner stops.

To maintain hydraulic bending machines you need to:

- check oil reservoir levels and top up with the correct hydraulic oil (never overfill)
- clean and lubricate all moving parts
- check the condition of formers and stop pins.

Pipe vices

These are portable devices that are used for holding LCS tube for cutting and threading. You need to use them with care, especially when using long lengths of tube that others may walk into or additional support is required at the end of the tube to prevent the vice falling over.

Notice how the jaws of the vices are curved and designed to hold the tube without causing distortion. Before attempting to thread tube, you should always check that the tube is secure in the pipe vice jaws.

To maintain pipe vices you need to:

- ensure that the vice jaws are in good condition (and replace if they are badly worn)
- check that the jaws can be wound down easily
- check that the folding foot stands are in good condition and the folding retaining device is positioned correctly.

TOOLBOX TALK

You need to take particular care when using the folding vices shown in Fig. 8.18. When transporting and erecting, ensure that the heavy folding foot stand is secure, otherwise it may fall and cause foot injuries.

Fig. 8.18 Pipe vices for LCS tube

ACTIVITY

While working on a building site, you are required to measure, cut and thread a length of 1″ LCS pipe 1500 mm long, which is to be used in a heating system.

List all the tools and equipment you would need to carry out this work safely.

Heating and ventilating materials and components

A vast range of materials and components are used within heating and ventilation, including boilers, flues and radiators, and heaters of various shapes, sizes and outputs. For this section of the chapter you will look at pipework, fittings and fixings. By the end of the section you should be familiar with:

- uses for LCS tube
- identifying pipe, fittings, valves and fixings for LCS tube
- prefabrication methods used to measure, cut, bend and install LCS pipework
- jointing methods used for LCS pipework
- testing LCS pipework installations.

DID YOU KNOW

The most common pipework materials are opper (see the plumbing section) or LCS.

LCS tube

LCS tube is rigid, strong and resilient. It is made from a mixture of iron and carbon, with the carbon content varying from 0.15–1.4%. It is manufactured to BS 1387:1985, which is the specification for screwed and socketed steel tube used for screwed or welded joints.

LCS tube is used for many applications, such as:

- heating systems
- gas pipework

- oil supply pipework
- compressed air systems (galvanised LCS tube)
- fire sprinkler systems
- hot and cold water supplies (galvanised LCS tube).

Due to this variety of uses, LCS tube is available in three different grades (see table below), with different weights, performance and cost. The grades are distinguished by a coloured band coding system and can also be identified by the difference in weight and the wall thickness of the tube.

Tube diameters are available from 6 mm (¼") up to 150 mm (6"), in lengths of 3.25 m or 6.5 m. Each length of tube has a black or red oxide protective paint finish with a BSP tapered thread on each end and one socket supplied.

Grades of LCS tube

Tube grade/duty	Colour-coding band
Light	Brown
Medium	Blue
Heavy	Red

Fittings for LCS tube

The most common fittings for LCS pipework are made of malleable iron, a form of white cast iron made when iron reacts with carbon to produce iron carbide, which is then **annealed**. It is manufactured to BS EN 10242:1995, which is the standard for threaded pipe fittings in malleable cast iron.

LCS tube used on air or domestic hot or cold water supplies has a protective zinc coating. This prevents corrosion and is referred to as galvanised tube or pipe.

TRADE TIP

Light-duty LCS tube walls are too thin to enable bending. If you attempt to bend the tube with hydraulic benders, the wall of the tube will collapse.

KEY TERMS

Anneal: A heating and cooling process that alters a material's properties.

Fig. 8.19 Malleable iron socket (showing reinforcing band) and a steel socket

Fig. 8.20 Pipe fittings

Malleable iron is able to accommodate some of the distortion that can occur with LCS tube. You can recognise it by the reinforcing band around the entry to the fitting, which reduces the chances of splitting the fitting when assembled. The internal BSP threads can be either parallel or tapered, depending on the manufacturer (parallel is the type most widely used). Fittings are also available in steel, which need no reinforcing band.

There is a vast range of fittings for LCS tube, including:

- sockets
- bends
- elbows
- tees
- bushes
- hexagon and barrel nipples (short pieces of tube used where fittings are very close together)
- unions
- long screw connectors
- flanges.

Fittings are available where all outlets are the same size or some have been reduced. Bends, elbows and branch outlets on tees are available in a range of angles.

Fitting outlet sizes can be reduced by using reducing bushes.

TRADE TIP

Unions, long screw connectors and flanges are all fittings to enable the assembly or disconnection of appliances, components or sections of a system.

ACTIVITY

Look at a manufacturer's catalogue (available online) to see the range of fittings available.

What is the difference between a bend and an elbow?

QUICK QUIZ

Which is the most commonly formed threaded joint for LCS tube and malleable iron fittings?

a) Taper to taper
b) Parallel to parallel
c) Taper to parallel

DID YOU KNOW

Bends with one 'male' end and one 'female' end are referred to as M&F bends. The male end of the fitting has an external thread and the female end has an internal thread.

Valves and other components

These are normally made of a different material to the pipe fittings, such as brass, bronze or gunmetal (an alloy of copper, zinc and tin), as they require more accurate engineering for the internal components.

You must be aware of the basic components used in heating and ventilating systems, such as:

- gate valves
- drain-off cocks
- air vents
- pipe clips and brackets.

DID YOU KNOW

The alternative to threaded fittings for LCS pipework is either to use a compression-type fitting or welded joints.

Gate valves

These are used to isolate sections of a system or appliances. They are often referred to as wheelhead or fullway valves, as they give very little resistance to the flow of water through the valve. They operate by turning the wheelhead connected to a spindle, which lifts a gate inside the valve to let water through. They must be fitted where they are accessible and can be operated and serviced.

Fig. 8.21 Gate valve (cut away to show workings)

Gate valves are also available as lockshield valves, which have no wheelhead. The valve spindle is enclosed to prevent accidental closure of the valve, and requires a key to operate.

On large systems, gate valves may have flanged joints so they can be easily replaced.

Drain-off valve

These valves are used to drain down the system or appliances once isolated. They are fitted in the system at the lowest point. The outlet side of the drain-off valve is designed to have a hosepipe fitted.

The valves are available as light- or heavy-duty pattern, and come with a lockshield cover to prevent accidental opening. They require a spanner or key to be opened. Drain valves are often prone to leaking through the spindle when open.

Fig. 8.22 Drain-off valves

Always ensure the hosepipe is secure when you are draining systems.

Air vents

Air vents enable air to be released from systems, and are usually fitted at the highest point. They are also fitted to radiators and to high points on sections of pipework.

Automatic air release valves, which open and release air from the system and then close once water is present, can be used, but are notorious for leaking.

Fig. 8.23 Air vents

Pipe clips and brackets

A wide range of clips and brackets are available for fixing LCS pipe, either directly to solid surfaces or suspended from walls or ceilings. They are made from brass or malleable iron.

The most commonly used clips for wall fixings, which allow you to space tube away from the wall, are:

- school board clips – these require countersunk screws
- munsen ring brackets – these come in three parts: base plate, clip and connecting rod, which can be purchased or cut to length (to enable various spacing distances from the wall)
- holderbat brackets – these are built into the wall, usually during construction.

Fig. 8.24 LCS munsen ring bracket and school board clip

At low level or where mechanical damage could occur, pipework should be clipped more often.

Spacing distances for LCS pipe clips

Pipe size	Internal vertical run	Internal horizontal run
15mm (½")	1.8m	1.8m
20mm (¾")	3.0m	2.4m
25mm (1")	3.0m	2.4m
32mm (1¼")	3.0m	3.0m
40mm (1½")	3.6m	3.6m
50mm (2")	3.6m	3.6m

REMEMBER

LSC pipe is very heavy and must be supported with clips that are well fixed.

Fixing devices for clips and brackets

The surface that you are fixing pipework to will determine the best type of fixing you should use (see page 252 on screws, raw plugs, expanding bolts and cavity wall fixings).

Basic installation of LCS pipework

Before you install LCS pipework, you need to know:

- exactly what is required
- where and how it is to be installed
- what it is expected to do once complete.

To ensure there is no confusion, two documents are provided – the specification and the plans (or drawings).

Typical drawings and specifications

The specification states:

- what the system is
- the type, make and size of the materials to be used
- what the performance criteria is and the standards that need to be achieved
- how the system is to be tested
- any specific safety considerations
- any other relevant information, including the manufacturer's instructions or literature.

Specifications vary depending on the complexity of the work.

The plans may be in various forms, such as:

- a block plan (this shows the position of buildings and structures on a site with scale of 1 : 2500)
- an elevation plan (used to show the height of buildings)

- a scale drawing (typically scales of 1 : 100, 1 : 50, 1 : 20 or 1 : 10)
- a line drawing (a simple drawing with dimensions)
- an isometric drawing (this enables three sides of an object to be viewed).

The drawings supplied to you will reflect the size and complexity of the work, but all drawings should give information on:

- the area in which the work is being carried out (e.g. part of a building or the dimensions of a work board for a small project)
- the layout of the system
- the diameter of the pipework
- the positions and dimensions of the pipework and components.

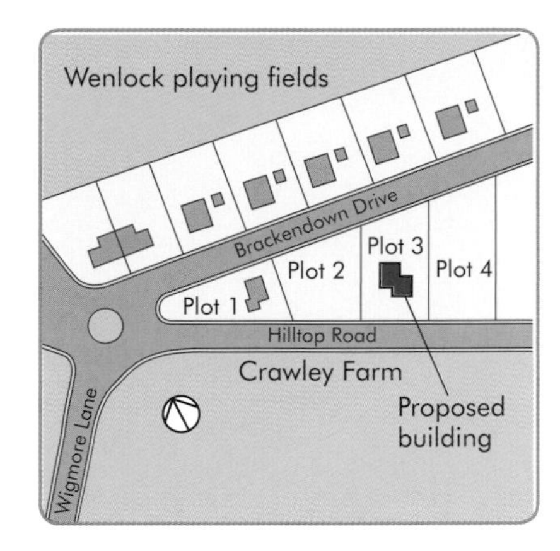

Fig. 8.25 A block plan shows the position of buildings and structures on a site

TRADE TIP

It is vital that you refer to specifications and plans before work starts. They will enable you to work out the job requirements, for example the fittings and length of tube required.

HAVE A GO...

Try drawing a simple object such as an oblong box (it can be to any dimensions).

Then make a plan of the box along with an end elevation and an isometric drawing complete with dimensions.

Measuring and marking out pipework

This first stage of the work is crucial if the specification criteria are to be achieved.

After you have ensured that all the documents, tools, equipment and materials you need are available and the area is safe to work in, follow these steps:

1. Mark out the position of the pipework and components to be installed, ensuring that all levels are correct.
2. Mark out the position of the clips, to ensure the pipework is adequately supported. You should also check that the clip positions will not coincide with any fitting or component positions.
3. Calculate the pipe lengths to be cut and threaded.

Z measurement method

The skill of calculating pipe lengths can seem difficult at first, especially considering that you have to work accurately to tolerances of plus or minus 3 mm. However, there is a simple, accurate method of calculating pipe lengths between fittings that is referred to as the **Z measurement** method.

Z measurement: The distance between the centre of the fitting and the end of the tube when inserted.

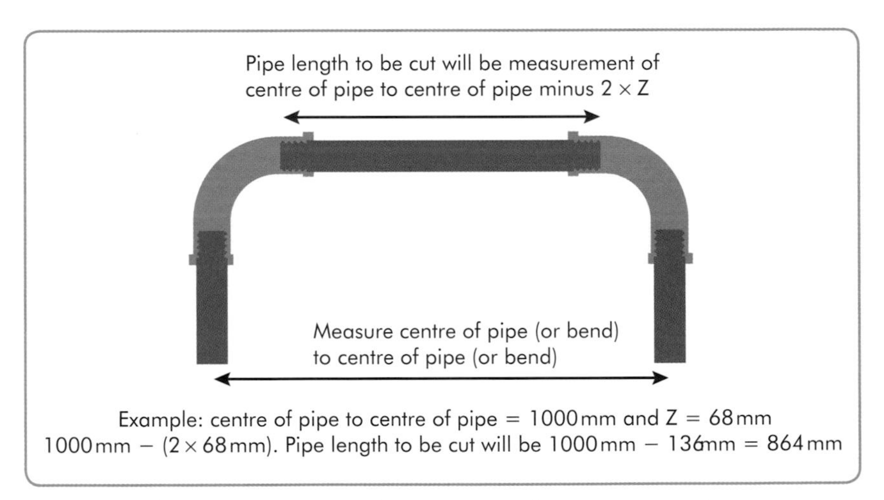

Fig. 8.26 Measuring and cutting LCS pipe using the Z measurement method

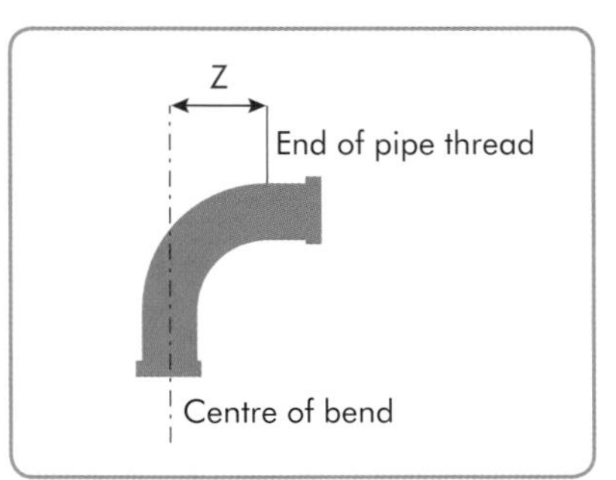

Fig. 8.27 Z measurement on a bend

The procedure is to either to look up the Z measurement in the technical table of the fitting manufacturer's catalogue for the fittings you are using or to measure this distance on the fitting to be used.

For example, if the distance between two 25 mm (1") bends is 1000 mm from centre of bend to centre of bend:

1. Calculate the Z measurement of one bend; for this example it is 68 mm.
2. Double the Z measurement (this now allows for both bends): 68 mm × 2 = 136 mm.
3. Subtract this measurement from the 1000 mm distance required between the two bends: 1000 − 136 = 864 mm.
4. The pipe can now be cut to exactly 864 mm long and, when assembled, the centre-to-centre distance between the two bends will be 1000 mm.

REMEMBER

The Z measurement will be different for each pipe size and every fitting.

DID YOU KNOW

Measuring, cutting, threading and assembling pipework ready for installation is referred to as prefabrication work.

Bending LCS pipe

Pipe is normally bent by using a hydraulic bending machine (see pages 215–16).

Creating a 90° bend from a fixed point is a simple operation if the following procedure is used:

1. Measure the distance from the fixed point (e.g. from the end of the thread into a fitting) to the position where the centre of the pipe needs to be when bent.
2. From this measurement, subtract the internal diameter of the tube (the pipe will gain this length when bent). For example, if 400 mm end of thread to centre of bend is required on a 25 mm pipe: 400 − 25 = 375 mm from end of pipe to centre of bend.

ACTIVITY

Using two 15 mm (½") elbows at 1000 mm centre to centre, calculate the length of the pipe that needs cutting.

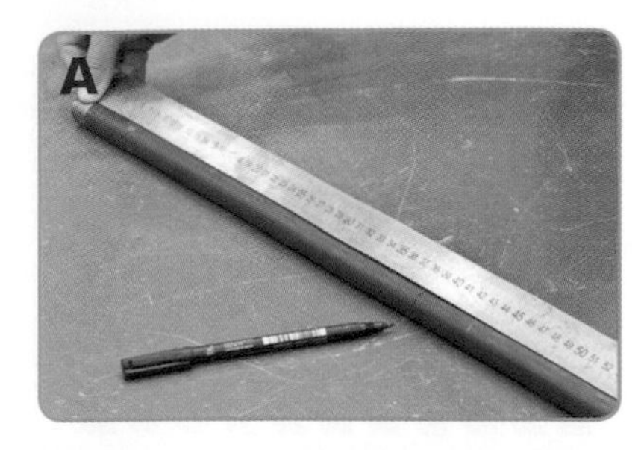

3. Transfer this measurement from the end of the pipe to be bent, and clearly mark the pipe (**A**).
4. Make sure the correct size former and stops are fitted (**B**).

5. Place the tube in the former with the mark on the pipe, in line with the centre of the former.
6. The bender is now pumped and the former will extend and push the pipe to the pipe stops (**C**).
7. Ensure the mark on the pipe is still in line with the centre of the former and continue to pump – this will apply more pressure and the bend will form (**D**).

8. Check the angle of the bend to ensure it is 90° (**E**) and then bend another 5° (you need to do this as the pipe will spring back by 5° when the pressure from the machine is released).
9. Release the pressure and pump the handle to return the former to its original position.

10. Remove the pipe from the former. (If the pipe sticks in the former, to release it do not hammer it out but remove the former with the pipe and then, holding the former, strike the end of the pipe onto a block of wood.) Check the bend is 90°.

Offsets

Making offsets with the hydraulic benders is carried out using a similar process:

1. Work out the offset required.
2. Mark the position required for the first offset (do not subtract any gain allowance).

3. Bend to the required angle (remember springback) (**A**).
4. Remove from the former.
5. Mark the position of the second offset. To do this, place a piece of timber along the back of the pipe past the first offset and then measure the offset distance required from the bottom of the timber to the pipe and mark it on the pipe (**B**).
6. Fit the pipe in the former, ensuring that the mark is in line with the centre of the former.
7. Check the pipe is level in the former (**C**) (or the offset will be twisted).
8. Pump the machine to form bend (**D**) (this should be 5° over to allow for springback) and then remove.

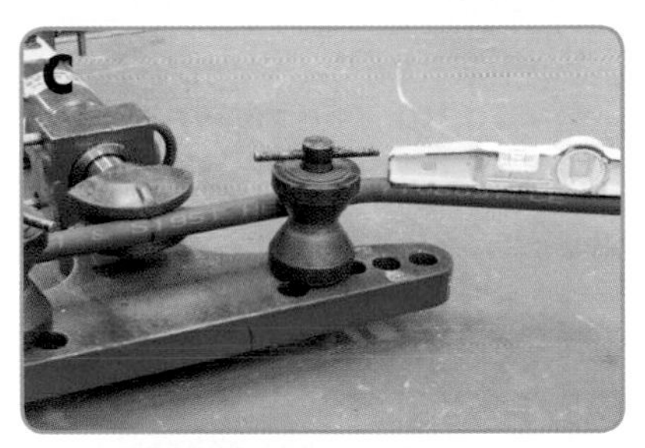

Threading and jointing LCS pipe

The procedure for cutting, threading and jointing LCS pipe is as follows:

1. Check the work area.
2. Make sure you have the documentation and authority to carry out the work.
3. Assess the work to be carried out and the materials required.
4. Make sure you have all the tools, equipment and materials available.
5. Set out the work and mark the position of any pipework, components and clips.
6. Fix clips and components as required (e.g. a radiator).
7. Measure the pipework lengths required.

REMEMBER

When forming 90° bends, the pipe will gain in length the same distance as the internal diameter of the tube being bent.

TRADE TIP

Incorrect positioning or fitting of pipe into the former can lead to rippling of the pipe while forming bends.

REMEMBER

When cutting and jointing LCS pipe, you should use the appropriate PPE in an area that has enough space for you to work without restriction and is free from hazards.

TRADE TIP

If dies are difficult to start when cutting a thread, file the edge of the pipe to form a slight angle (chamfer). This will give cutting dies a better start.

DID YOU KNOW

PTFE stands for polytetrafluoro-ethylene.

8. Fit the pipe into the pipe vice.
9. Cut the pipe to length.
   - If cutting with a hacksaw, file inside and outside and ensure the cut is made square, then remove swarf from inside the tube.
   - If cutting with pipe cutters, ream out the internal bore of the pipe.
10. Form any bends or offsets required.
11. Thread the ends of tube using stocks and dies and making sure that:
    - the pipe is securely held in the pipe vice
    - adjustable dies are correctly set
    - thread lengths are cut two threads longer than the length of thread inside the fitting
    - a means of catching any surplus thread cutting oil is placed below the pipe (as the pipe needs to be lubricated with thread cutting oil during cutting of thread to avoid damage to both dies and thread).
12. Clean off surplus oil and remove from inside the pipe.
13. Repeat this process as necessary for each length of pipe to be cut.

Installing LCS pipe

Once the pipes have been cut and threaded, the joints can be assembled. The procedure for this is as follows:

1. Make sure the pipe is firmly held in a vice.
2. Apply jointing compound or PTFE tape to the thread. The type of jointing compound you use will depend

on the system – it must be suitable and approved for use with the system's contents.

3. Carry out a visual check on the fitting for:
   - blockages
   - condition of internal threads
   - pin holes in castings
   - any other deformities or excess distortion.
4. Hand-tighten the fitting onto the thread.
5. Use a pipe wrench to tighten the fitting onto the thread to form a seal. Do not over-tighten as this may damage the fitting.
6. Remove any excess jointing materials.
7. Repeat the process as necessary.

Depending on the job at hand, it may be possible to prefabricate the work before it is fitted to the system or structure. However, sometimes it will be necessary for you to install the pipework as the work proceeds.

There will be occasions where a union-type connection or flanged joint will need to be fitted. These are often needed to facilitate the installation of pipework, or when removal of pipework or components for service or repair is required.

Tapered joint unions consist of three parts: one section that fits to the end of each pipe to be connected and a nut that pulls these together and makes a seal with a tapered joint.

Fig. 8.28 LCS threaded union joint

REMEMBER

You need to take care to avoid excessive marking of pipe and fittings when tightening joints with a pipe wrench.

TRADE TIP

It is often easier to tighten fittings onto pipe by using a short length of pipe in the fitting as a lever.

Fig. 8.29 Flanged joint (top) and a union (below) installed in LCS pipework

Other unions and flanged joints are flat-faced fittings that require a sealing washer or gasket to make a seal between the two faces of the fitting to be joined.

It is important that you correctly align both unions and the flanged joint before tightening. Most flanged joints have four holes in which nuts and bolts are used to tighten the two flange faces together (all four bolts should be tightened to the same pressure).

Commissioning and testing systems

The completed installation must be tested before it is used.

Test procedures will vary depending on the type of system or pipework installed. For example, hot and cold water supply systems have to be tested according to the procedures in BS 6700:1987, which is the specification for the design, installation, testing and maintenance of services supplying water for domestic use within buildings and their curtilages. Generally, however, systems are tested to 1.5 times their normal working pressure for a specified length of time.

When testing heating and ventilation system pipework, there are two methods that are normally used:

- Hydraulic pressure testing – air and water are pumped into the system up to the pressure specified. This method has the advantage that if the system leaks within a building then the leak is easier to locate (usually seen as a damp patch near the leak or heard as the air leaking out under pressure). Also, as only air and a small amount of water are used, there is less chance of any leaks causing damage to the building structure.

- Testing by filling the system with water – this is a less suitable method, as there is potential for damage to the building if leaks occur.

Test procedures and pressures are often detailed in the specification for the work to be carried out.

Fig. 8.30 Hydraulic pressure-testing equipment

For either test the procedure is similar:

1. Carry out visual checks of all pipework, components and appliances to ensure that:
   - all pipe joints have been completed
   - nuts to valves and fittings have been tightened
   - all taps and pipe ends are closed or sealed
   - all drain valves and air vents are closed.
2. Make sure that anyone in the area of the testing is aware that testing is being carried out.

REMEMBER

You can only pressure test a system up to the pressure of its weakest component.

3. Connect the pressure-testing equipment to the system to be tested, ensuring that you have a pressure gauge.
4. Pressurise the system and test for a specified time.
5. Check the gauge during the specified time to ensure the pressure has not dropped through leaks.
6. Carry out visual checks of the system during the test, to check for leaks.
7. If the system passes the test after the time specified, release the pressure and remove and test the equipment. If the system pressure drops, then you will need to locate the leak, repair it and re-test the system.
8. Thoroughly flush out the system, to remove debris before filling.
9. If the system is for hot water or heating pipework, heat the system, allow it to cool and then check for leaks again.
10. Take notes of all test results.
11. Test and adjust any system controls.
12. Clean the site or work area and tidy away tools and equipment.
13. Hand over to the client and inform them about use and operation of the system.

DID YOU KNOW

Systems may pass a test with air or cold water but leak when hot water causes expansion and contraction of pipes and fittings.

CHECK YOUR KNOWLEDGE

LEVEL 1

1. What card is required to work on a building site?
2. What type of hacksaw blade is used to cut LCS pipe?
3. What is a reamer used for?
4. Name two types of stocks and dies.
5. What other documents are used with plans?
6. Name two types of clip used to support LCS pipe.
7. List the basic process for assembling a threaded joint for LCS pipe.
8. What two methods could be used to test pipework for leaks?

9 Plumbing operations

This chapter covers the learning outcomes for:

Understand and demonstrate fundamental plumbing operations

City & Guilds unit number 108; EAL unit code AQCC1/08; ABC unit code A07 for Level 1 and A05 for Level 2.

Plumbing describes the work a plumber does. This includes working with pipes and fixtures that carry or use water from the incoming mains supply all the way to the sewer.

Every house or building, including where you live and study, will have some form of plumbing. As you study this chapter, try to identify in your surroundings any new aspects of plumbing you learn about.

In this chapter, you will learn basic plumbing tasks. You will also discover the type of tools, materials and equipment a plumber uses in their work.

IN THIS CHAPTER YOU WILL LEARN ABOUT:

- health and safety methods when carrying out basic plumbing work
- plumbing hand tools
- plumbing materials and components
- basic plumbing practical applications.

Basic systems

As a plumber, you will come across four basic systems:

- Cold water
- Hot water
- Heating
- Sanitation

Cold water systems

Direct cold water systems are the most common. They are a cost-effective installation and are usually appropriate because:

- of the high pressure of supply available
- less pipework is required
- it is possible to draw drinking water from numerous points
- there is no need for a cold water cistern if the hot water is also direct.

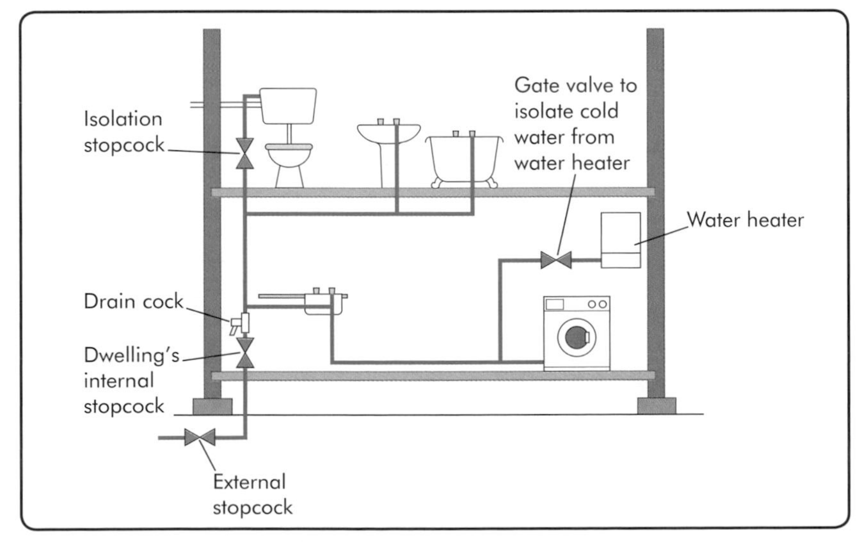

Fig. 9.1 A direct cold water system

The only problems are that if there is no mains supply then there is no reserve of cold water. This means that taps and valves suffer greater wear and tear due to the high pressure created in the system.

Indirect cold water systems are used in low pressure areas. They incorporate a reserve of stored water, which becomes available if the mains supply is turned off. They usually require a minimum 230 l storage cistern and extra pipework.

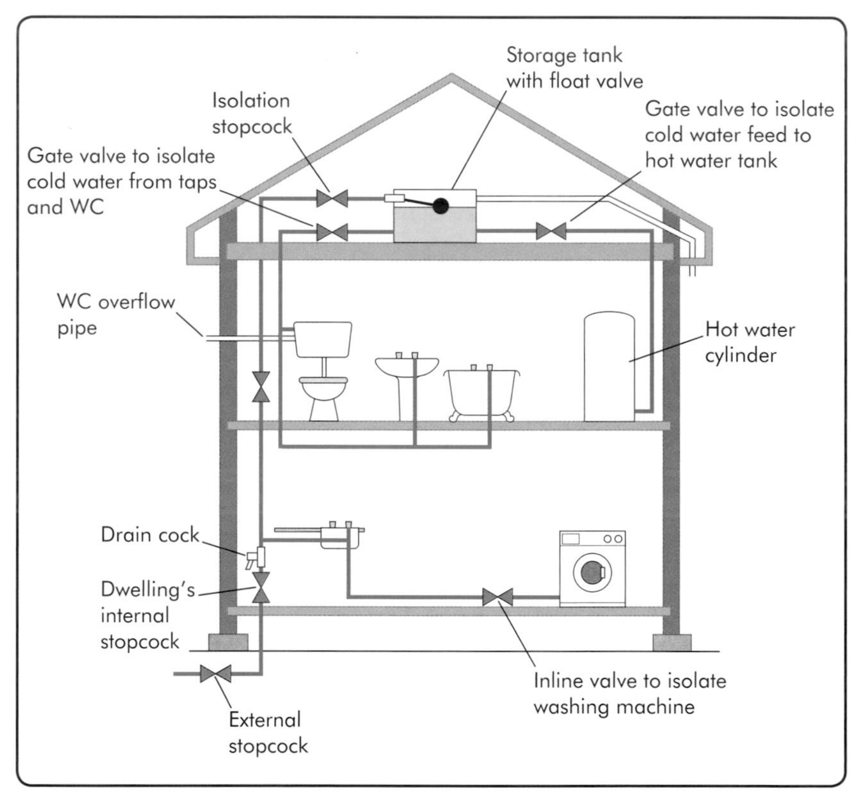

Fig. 9.2 An indirect cold water system

Heating systems

There are various options when installing heating systems, but most combine the heating of hot water and radiators.

The most popular system currently is the combi boiler. At the heart of this system is a combination boiler, which immediately heats cold water to be used as a direct hot water supply. It also supplies hot water for the central heating system.

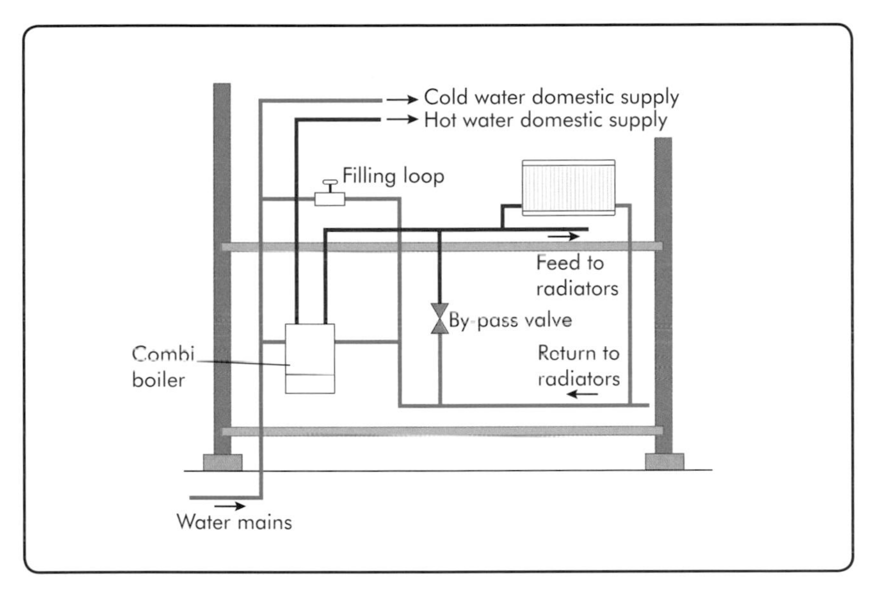

Fig. 9.3 A combination boiler heating system

Sanitation systems

The single stack sanitary or drainage system usually has a vent at the top of the soil pipe. Each bath, basin or WC has a U-bend to prevent sewer smells from escaping into the dwelling. The water acts as a barrier.

Single stack systems need to be carefully designed so that there is no likelihood of smells entering the building. Note the manhole inspection cover in Fig. 9.4, to allow access for dislodging blockages.

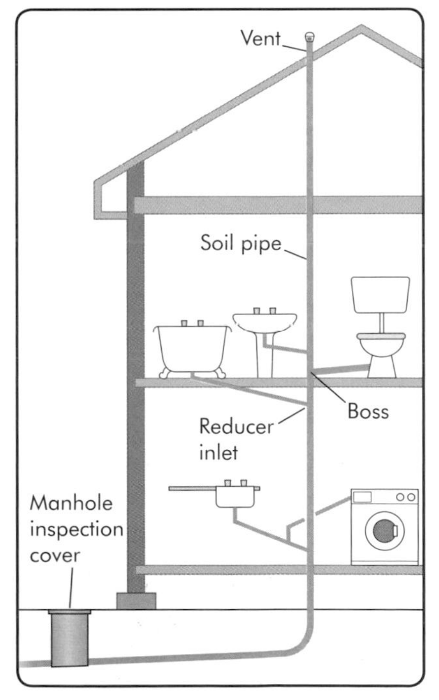

Fig. 9.4 A single stack sanitary system

Health and safety methods when carrying out basic plumbing work

DID YOU KNOW

Inexperienced workers are particularly at risk when carrying out plumbing work.

When carrying out any practical tasks, you need to think carefully about health and safety. Whether you are in the workshop or completing a job outdoors, you are required by law to adhere to all health and safety regulations. In practice, this means that you should take reasonable steps to ensure your own health and safety and that of anyone who may be affected by what you are doing.

Before any job is undertaken, a risk assessment must be carried out based on:

- the work you are about to do
- the environment in which you will carry out this work
- the materials and substances you are about to use.

Personal protective equipment

The personal protective equipment (PPE) that you wear to carry out plumbing work includes:

- safety boots
- overalls
- safety helmet
- safety goggles
- gloves
- ear defenders.

Fig. 9.5 Safety boots

All plumbing tasks that are undertaken can be a risk, not only for the person carrying out the task but also to anyone else around. This means it is important to make sure that your surroundings are kept hazard-free and don't pose a risk to others. These are some points to remember:

- When carrying out plumbing activities, try to keep your workspace neat and tidy.
- Don't leave tools and materials lying around, as they could become a trip hazard.
- If the work you are doing creates dust or fumes, make sure it is carried out in a well-ventilated area.
- If you are using electrical tools, take care not to leave trailing leads, as they also can become a trip hazard.
- When using a blowlamp, to prevent the risk of fire, it is essential that you protect any combustible material.

TOOLBOX TALK

When working with waste pipes, always wear protective gloves. These will protect you from Weil's disease, which is spread by rats.

ACTIVITY

Which PPE would be required for the following tasks? (You may use more than one per task.)

- Cutting a piece of metal with a hacksaw
- Unblocking a sewer pipe with cleaning rods
- Drilling a hole into brickwork
- Carrying out a task where others may be working above you

REMEMBER

If you see any potential workshop hazards, you must report them. This is your duty by law.

TOOLBOX TALK

When using a blowlamp, always have a fire extinguisher close at hand (see page 61 for extinguisher types).

Plumbing hand tools

When carrying out plumbing activities, you will be using an assortment of different tools. Some tools are generally used across all trades, while others are specific to the plumbing industry.

In this section, you will learn about a selection of hand tools that a plumber uses.

Hacksaw and junior hacksaw

TRADE TIP

Some form of lubricant on the hacksaw blade will ensure a smooth cut and longer blade life.

A hacksaw is used for general cutting of metal and plastic, including pipes. The blade can have 18, 24 or 32 teeth per 25 mm. You will usually find that the softer the material being cut, the fewer teeth required on the blade.

A junior hacksaw will always be included in a plumber's tool kit. It is smaller than an ordinary hacksaw, requiring only one hand in its use.

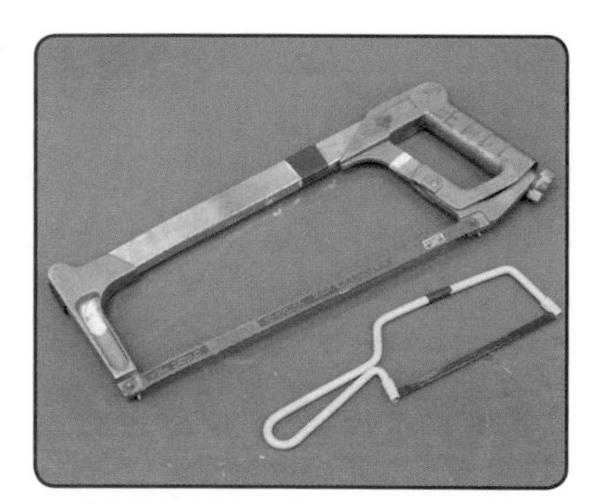

Fig. 9.6 Hacksaw and junior hacksaw

When using any form of hacksaw, you need to make sure that the blade is sharp with no broken teeth. You should also check that the teeth face forward.

When cutting, apply pressure on the forward stroke as this is when the material is being cut.

Pipe cutters

Pipe cutters can be used to give a neat, finished cut to pipes constructed from different types of material, including copper, plastic and iron. There are a number of different types of cutters, some of which are described below.

Roller cutter

Roller cutters use a cutting wheel and rollers. They incorporate a form of tightening handle that is turned as the cutter is spun around the pipe. They come in various sizes in accordance with the size of pipe being cut.

Roller cutters have a tendency to squeeze the end of the pipe, so reducing the internal diameter. You will therefore need to deburr the pipe once cut, using either a burring reamer or a half-round file.

DID YOU KNOW

Larger roller cutters can also be used to cut steel pipe.

This type of cutter also needs regular maintenance, as dirt and paint is picked up on the rollers when cutting and can make them seize up. To avoid this, roller cutters should be routinely dismantled and cleaned.

The pins that hold the rollers and cutting wheel also tend to get worn, the result of which is called 'tracking'. Tracking appears as two or more cuts along the pipe, rather than actually cutting through it.

TRADE TIP

Be sure to remove any burrs when cutting pipe, as they could damage the pipe or snag any residues. If this happens, the burrs will cause a blockage and prevent water flow.

Plastic pipe cutter

This is another type of cutter that you will use in the workshop. It has a scissor-type cutting action on a ratchet and very sharp blades. It squeezes and locks onto the pipe as you squeeze the handles. You need to be very careful with this cutter as it can easily slice off your finger.

Before using, it is important to check that the blade is in good condition with no broken edges.

Little maintenance for these cutters is required, except for some lubrication occasionally and blade replacement, which is only possible with some brands.

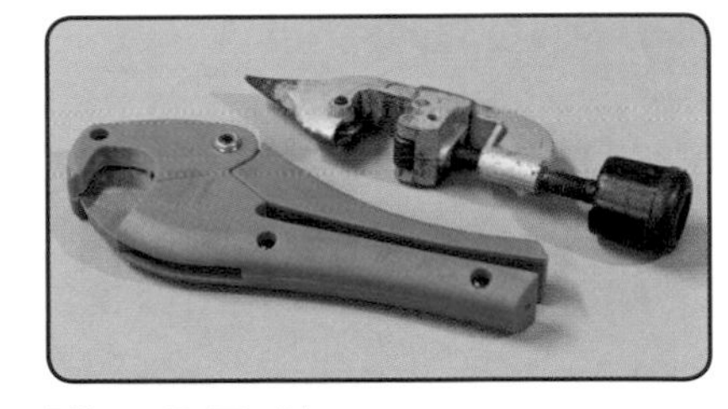

Fig. 9.7 Plastic pipe cutter and roller cutter

Adjustable spanner

An adjustable spanner is a plumber's best friend as it takes away the need to carry different-sized fixed-head spanners and it is used to tighten all sorts of compression fittings.

Fig. 9.8 Adjustable spanners

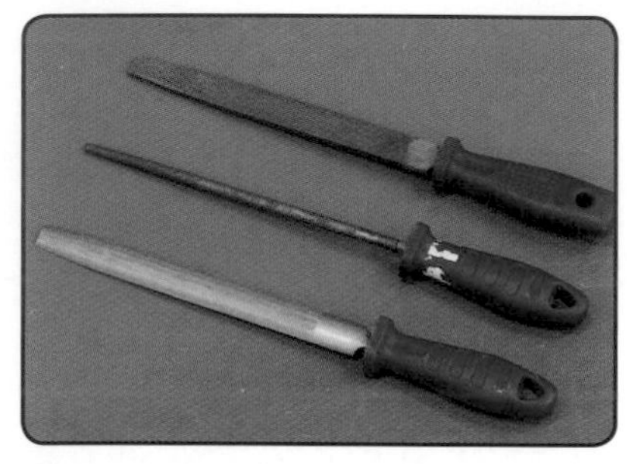

Fig. 9.9 Files

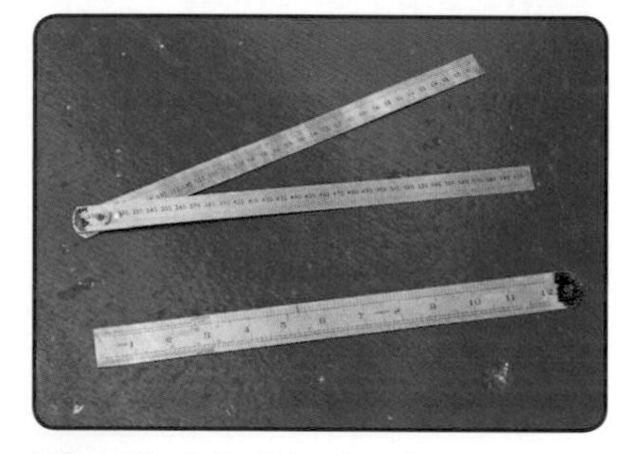

Fig. 9.10 Steel rulers

TRADE TIP

Tape measures and steel rulers give dimensions in imperial and metric. However, to get accurate measurements we tend to use metric, more specifically millimetres.

When using adjustable spanners, be sure to tighten them as much as possible around the nut of the compression fitting. This is because they can slip off if too loose and cause damage to the building fabric or a nasty injury to the user.

Adjustable spanners require regular lubrication and cleaning – otherwise they tend to seize up.

File

A file can come in various shapes and sizes. It is mainly used to file down any rough edges and especially to deburr pipes that have been cut with a pipe cutter.

Never use a file with a damaged or improper handle as this could cause injury to the user by puncturing the skin.

No maintenance is usually required for files. If the handle does break then the file can no longer be used and you will need to get a new one.

Measuring equipment

A tape measure is essential in a plumber's toolkit. You will need it to measure and cut pipe accurately and for other plumbing activities.

A steel ruler is used when smaller and finer measurements are required. This is especially the case when completing jobs in the workshop.

Tape measures are self-closing on a spring action, so care should be taken that they don't recoil too quickly as they could catch your skin. It also shortens the life of the tape measure as it damages the spring.

HAVE A GO...

Cut a piece of copper 200 mm in length using pipe cutters.

Be sure to deburr the ends once cut.

Screwdrivers

Screwdrivers can be flathead or crosshead, depending on the type of fixing used. You will need them when fitting clips and brackets for pipe.

Screwdrivers should never be used as chisels. Apart from the fact that this could ruin them, it is also dangerous – the handles are not built to withstand impact and could break and cause the user injury.

The only type of maintenance required is an occasional wipe clean of the handle.

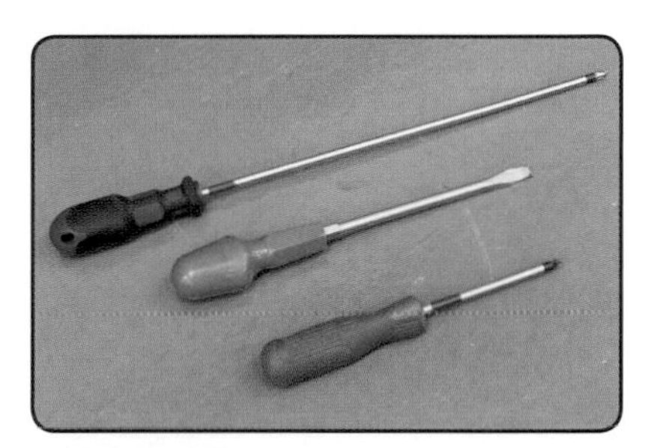

Fig. 9.11 Flat and crosshead screwdrivers

TOOLBOX TALK

Always use the right size screwdriver to match the fixing. The wrong size could slip off and cause injury.

Spirit level

A spirit level is needed to ensure that fixtures and fittings are installed level. You can buy them in various sizes, from 5 cm to 100 cm in length.

You need to take care to avoid heavy knocks to spirit levels, as this could affect the bubble tube and result in a false reading.

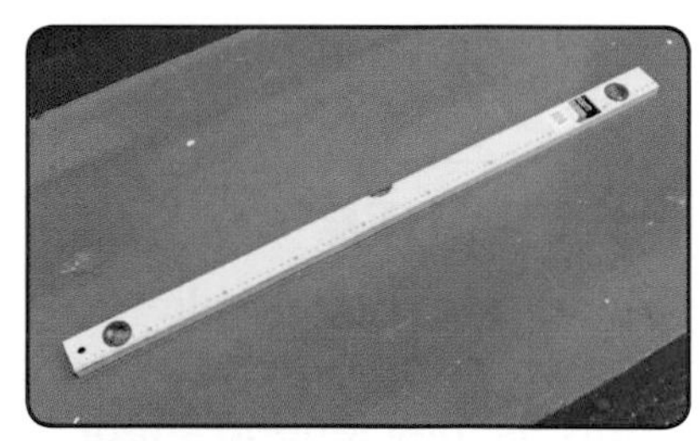

Fig. 9.12 Spirit level

Blowlamp

A blowlamp is used to solder copper fittings onto copper pipe. It usually burns propane gas.

Fig. 9.13 Blowlamp

TOOLBOX TALK

The nozzle of the blowlamp remains hot for a while after turning the blowlamp off, so be careful not to touch it!

If used incorrectly or carelessly, you could set fire to the fabric of the building. For this reason it is vital that you always use a heat mat to protect surrounding areas.

When soldering, take care not to apply too much solder to the joint. Excess solder will either run down the pipe, making an unsightly joint, or drip onto the floor and damage the floor covering.

Care must also be taken not to damage or clog up the nozzle, as this will impair the blowlamp's performance.

Water pump pliers

Water pump pliers are specifically designed to remove the compression fittings you find on a central heating pump. They are larger than normal pliers and are needed when adjustable spanners cannot be used. They can also be used for a variety of other jobs.

Regular lubrication and cleaning will ensure a longer life for your water pump pliers.

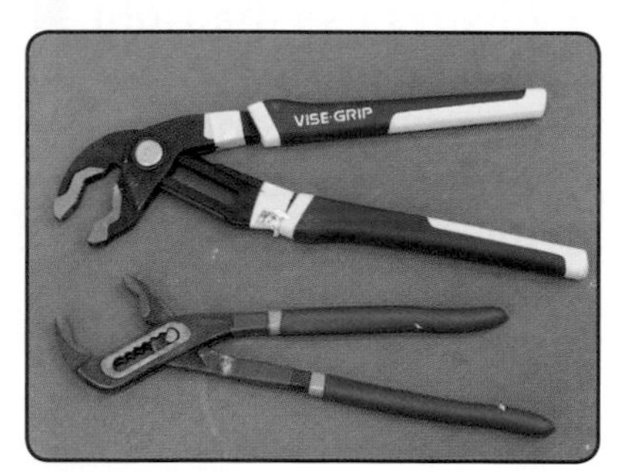

Fig. 9.14 Water pump pliers

REMEMBER

Avoid using pump pliers on ordinary compression nuts as they tend to shear the edge off the nut.

Bending springs

Bending springs are used to manually bend pipes. They can be internal or external, and are used to prevent the walls of the pipe from collapsing.

When using bending springs, you normally pull a bend slightly more than is required. This allows you to then pull the bend back in order to facilitate the removal of the spring.

When using bending springs:

- be careful not to pull too tight a radius, as it will be hard to remove the spring
- do not use damaged or distorted springs, as these will be hard to use
- regularly lubricate to assist in the insertion and extraction of the spring.

Pipe benders

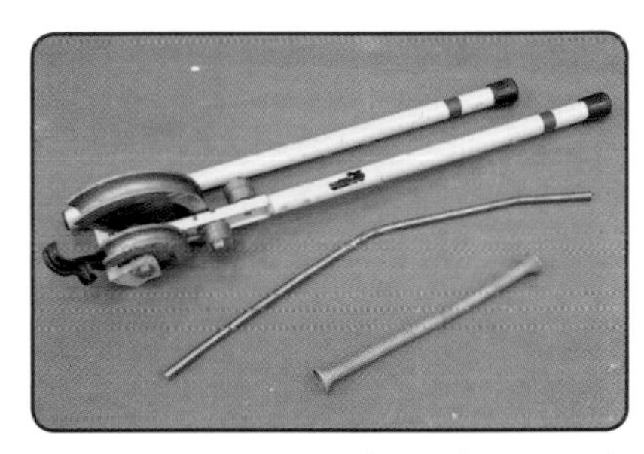

Fig. 9.15 Pipe benders and bending springs

Pipe benders are used to form bends in copper pipe. Doing this will save the need to use fittings.

Pipe benders work on the principle of leverage. They can be either hand-held or on a stand (for larger diameter pipe).

Hand-held benders, also called scissor benders, consist of two bend dies (formers) in which the pipe sits, which are normally bent at 90°– one for 15 mm tube and the other for 22 mm tube. A clamp die suitable for both pipe measurements holds the pipe in place, while rollers that assist in keeping the right pressure are applied. A pressure die for each pipe diameter, normally called a guide, is inserted between the pipe and the rollers.

It takes time to master the art of pipe bending, but once learnt it can save you time and money.

When bending with mechanical benders, the wall of the pipe stretches, so care has to be taken that the rollers are correctly tightened.

A bit of physical effort is required when using pipe benders, especially with 22 mm pipe, so you will need to be careful to avoid muscle injury or strain.

TRADE TIP

When replacing a hacksaw blade, make sure the cutting teeth are forward facing.

Regular maintenance of pipe benders is advisable, and particular attention should be given to ensuring the rollers and guide are kept clean.

ACTIVITY

Name these tools and state what they are used for.

A

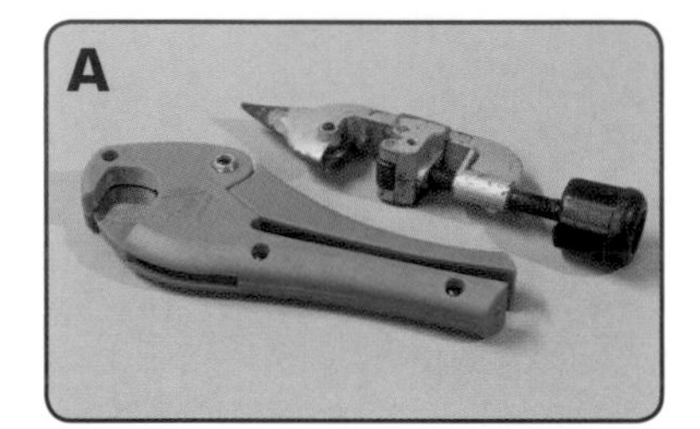

B

C

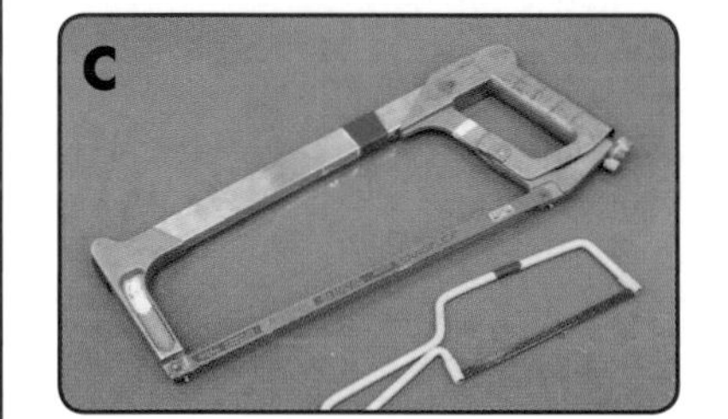

D

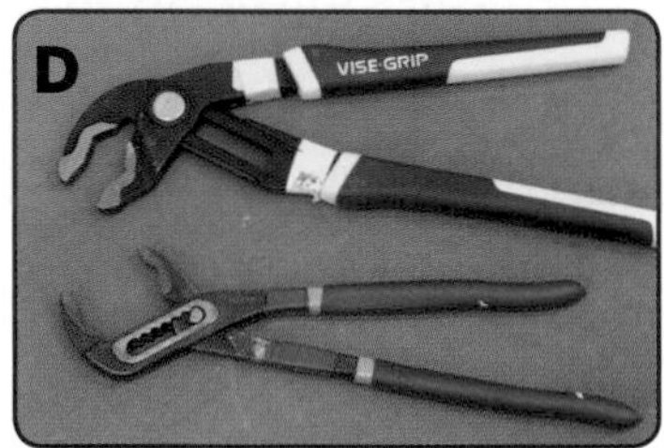

REMEMBER

Keep your toolbox neat and tidy, as this will save you time on the job when looking for tools.

DID YOU KNOW

The price of copper has tripled over the last 10 years. This makes plastic pipe a cheaper option.

Plumbing materials and components

There are various types of materials that can be used for plumbing activities, such as copper, plastic and steel, for the pipework that carries the water.

Plastic pipe is commonly used for cold water, hot water and central heating as it is easy to cut, bend and install. Plastic pipe is also quicker and cheaper to install than copper. The only drawback is that plastic pipe does not look as good as copper pipe, so it is mainly used where it is hidden.

Copper pipe, fittings and components

This section will concentrate on copper pipe and the fittings used to connect pipe together.

For most domestic plumbing applications, 15 mm or 22 mm diameter pipes are used. These normally come in 3 m lengths.

You can also get pipe in the following, less standard sizes: 6 mm, 8 mm, 10 mm and 12 mm. These sizes are mainly used for other trades, such as air refrigeration and gas work. The pipe is more **malleable** and usually comes in 25 m coils.

Copper tube is also available in sizes of 28 mm, 35 mm, 50 mm and 65 mm. These sizes of pipe are normally used in commercial applications, such as schools and factories.

You can join copper pipe together using a variety of methods and fittings.

- The simplest way is to use plastic push-fit fittings, which push onto the end of the pipe.
- A compression fitting is also popular and easy to use. This type of fitting uses a back nut and compression ring (**olive**) inserted on the end of the pipe, which is tightened using adjustable spanners to make a water-tight seal. PTFE or jointing compound is applied around the olive prior to tightening.
- The most common and cheapest fitting is a copper capillary fitting. This is welded onto the pipe by means of a soft **solder** using a blowlamp. This process is called soldering. It utilises **capillary action**

KEY TERMS

Malleable: Describes something that is flexible and can be bent without fracturing (breaking).

Olive: The more common name for the compression ring made from brass or copper that is used in a compression fitting.

Solder: Usually comes on a reel and is made from alloys of metal, such as tin or copper. It has a low melting point of about 450°C.

Fig. 9.17 A compression joint fitting

KEY TERMS

Capillary action: The process that causes liquids to rise up small gaps between two surfaces of solid material.

Fig. 9.18 Yorkshire fittings

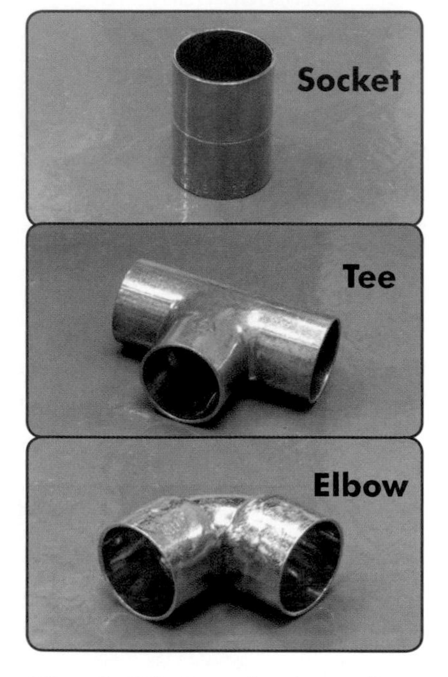

Fig. 9.19 Standard capillary fittings

in order for the solder to fill in the gap between the fitting and the pipe and so make a watertight seal.

Some fittings already have the solder incorporated, which make them more convenient. However, these types of fitting, called Yorkshire fittings, are more expensive than ordinary capillary fittings.

Copper pipe connects together with fittings. Different fittings are available:

- A socket is a type of coupling that is used to connect two bits of copper tube in a straight length.
- If you want to create a 90° bend with two bits of pipe then an elbow is required.
- In order to connect three bits of copper pipe at 90° separation, a tee is used.

You can also get fittings that create a 45° bend, and even one that creates a passover, which is used to go around another section of pipe.

HAVE A GO...

Make a compression joint using a socket, PTFE tape and a section of copper pipe.

ACTIVITY

Can you work out how many different fittings have been used in order to create the pipework shown?

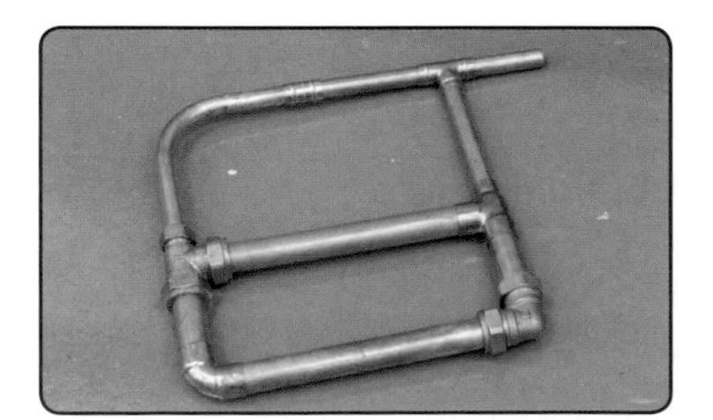

Fixing pipework

Fixing devices used to secure pipework are called clips. They should provide enough support to prevent accidental damage and movement of the pipe, which otherwise can create a knocking noise called water hammer.

British Standards have recommendations for spacing of clips that are determined by the construction and size of the pipe being supported.

Clips can be constructed from a variety of materials, including copper, steel, brass and plastic.

- Copper clips are called saddle clips as they clip the pipe back tightly to the wall by sitting over it like a saddle on a horse. This type of clip can cause a chemical reaction between the pipe and the wall construction, which can lead to the pipe corroding.
- Steel clips are only used for steel pipe.
- Brass clips are mainly used in commercial situations where a more secure fitting is required. They can be of a school board bracket type or a munson ring type, which fits onto either a threaded back plate or threaded bar.
- The most common clips used in domestic plumbing are plastic ones. These have the advantage of needing only one fixing screw in which the pipe 'snaps' into, so saving time,. These clips are also the most economical (cheapest) to purchase.

Other fixings used include holderbat and canterlever brackets, which are used in schools and offices and to support heavier fixtures. These are normally built into the wall using a strong mortar mix.

Fig. 9.20 Various clips and pipe fixings

TRADE TIP

Be careful when soldering pipes close to plastic clips, as the heat travels up the pipe and will melt them. In order to avoid this, you will either have to remove the clip temporarily or solder the pipework on a bench if possible.

DID YOU KNOW

Using steel clips on copper pipe will eventually corrode the copper by means of electrolytic corrosion (see pages 102–3).

Fig. 9.21 Screws and nails

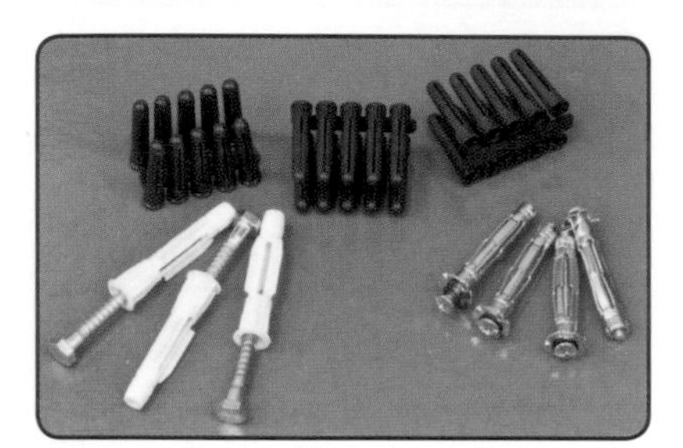

Fig. 9.22 Fixings

When fixing heavy fixtures, such as a radiator on to a hollow wall, wood noggins are required between the uprights to ensure a stable fixing(see Fig. 9.23).

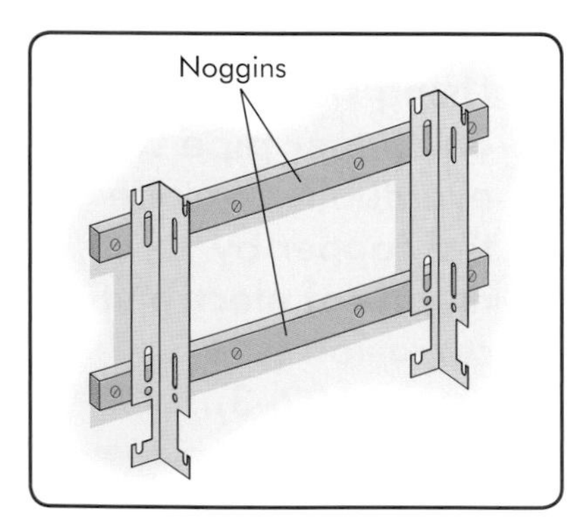

Fig. 9.23 Supportive noggins are needed on stud walls for heavier items such as radiators

Fixing devices

A variety of fixing devices are used in the plumbing industry, including the following:

- Wood screws, so named because they screw into wood, can have either a slotted head or crosshead. They can also be countersunk, mushroom or roundhead, depending on the fixture.
- When fixing into a solid wall (masonry), a raw plug constructed of plastic is the cheapest and most reliable fixing to use. It is inserted in to a pre-drilled hole of the correct dimension for the raw plug used with the screw, and then fixed into it.
- A spring toggle fixing is normally used for hollow blocks or steelwork. It enables the fixing load to be spread over a larger surface area.
- Expanding bolts are used in dense building materials, such as concrete, to secure heavy fixtures.

ACTIVITY

Choose the correct fixings for each wall construction below and state which type of clip you would use if required.

1. A 22 mm copper pipe is to be fixed onto a brick wall.
2. A 28 mm steel pipe is to be fixed onto a concrete floor.
3. A radiator is to be fitted onto a plasterboard/ hollow wall.
4. A 15 mm copper pipe is to be fitted onto a plasterboard/hollow wall.

Basic plumbing practical applications

Measuring and marking out for pipework

To carry out practical plumbing applications, you will need to be able to measure and mark out accurately. This is so you can cut or bend the pipe and then fabricate it according to specifications. A tape measure or steel ruler is used for this.

In the workshop, most applications will need to be constructed to within 2 mm of the **specification**, so you will need to be very accurate in your measurements.

KEY TERMS

Specification: A detailed, exact statement of particulars, describing materials, dimensions, and quality of work, for something to be built, installed, or manufactured.

Cutting pipework

Copper pipe is usually cut using pipe cutters, while plastic pipe cutters are used for plastic pipe. A junior hacksaw can also be used on smaller diameter pipe, but with this method it is harder to achieve a straight cut.

Whichever tool you use, you will need to measure the length of the section of pipe you require. Mark on the pipe using either a pencil or a permanent marker. It is important to be as accurate as possible, so do not rush your measuring and marking.

Using a roller cutter for copper pipe

1. Be sure to fit the pipe to be cut between the rollers and the cutting wheel, with your mark on the cutting wheel visible.

2. Turn the tightening handle until the pipe is firmly held in place between the rollers and the cutting wheel.
3. Hold the pipe with one hand while also rotating the cutters around the pipe.
4. Apply a slight pressure on the pipe with the cutting wheel by turning the tightening handle.
5. Be careful not to over-tighten the handle, as this would cause the cutting wheel to cut too deep into the pipe. If this happens, it would crush the wall and prevent the wheel travelling along the diameter of the pipe. This action can also cause damage to the cutting wheel.
6. Once the pipe has been cut, make sure you hold both sections of the pipe to avoid damaging the ends or snapping the cutting wheel.
7. Finally, remember to deburr the cut end.

Pipe bending

When fabricating pipe, it is cheaper and quicker to avoid soldering if possible. This can be achieved by bending the pipe using a spring or a pipe bender.

Pipe bending machines can be heavy, so take care when handling them.

A spring is mainly used for simple bends like 30° offsets in the pipe. It is especially useful when the pipe is in position and cannot be put into a bending machine or if you require a bend that cannot be formed with a pipe bender.

Competent plumbers will always have a pipe bender in their toolkits and will have mastered the techniques for bending copper pipe. This will save time and money and the finished work will look neat and tidy. Remember, a customer will be more agreeable to pay the bill if the work completed looks good.

Some points to remember when using a pipe bender:

- You can bend pipe to various angles, from small offsets at 30° and 45° to 90° bends and complicated passovers, as long as you have enough pipe to sit in the machine to bend.
- In order to achieve the correct bend required, you will need to measure accurately. This is normally done using a steel rule.
- When cutting pipe for bending, you will have to measure accurately, so as not to waste any pipe.
- The rollers on pipe benders should be adjusted to suit the guide and the pipe. If adjusted too tight, you will cause a 'throating' effect, which flattens the pipe on the bend, reducing the diameter of the pipe.
- Adjusting the rollers too loose will cause a ripple effect on the pipe wall on the bend, with raised edges making the pipe look ugly.

REMEMBER

When bending copper pipe to 90° or more, the pipe stretches making it longer.

Different bends have their own unique method of bending. We will now look at how to form a 90° bend.

Bending a pipe 90°

If, for example, you want to have a bend at 200 mm from the edge of 15 mm pipe to the centre of the bend, follow these steps:

1. Make a mark 200 mm less half the diameter of the pipe from the edge of the pipe (**A**). This distance 200 mm − (15/2) = 192.5 mm is called 'distance X'.

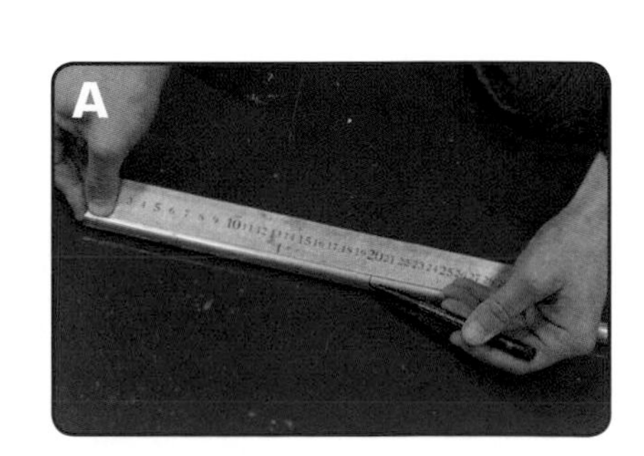

2. Insert the pipe into the pipe bender nestled into the former and held in place with the guide and clamp (**B**).
3. Using a square placed against the back of the former and under the pipe, move the pipe until the distance X mark is in line with the edge of the square. Alternatively, hold a pipe at right angles in the bender and make sure the line sits in the centre of the pipe (**C**).
4. Apply pressure to the lever arm and pull the bend to approximately 95°. The reason you slightly over-pull the bend is because it will return on itself a little (**D**).
5. Before releasing the guide and clamp, check with the square that the bend is at 90°.

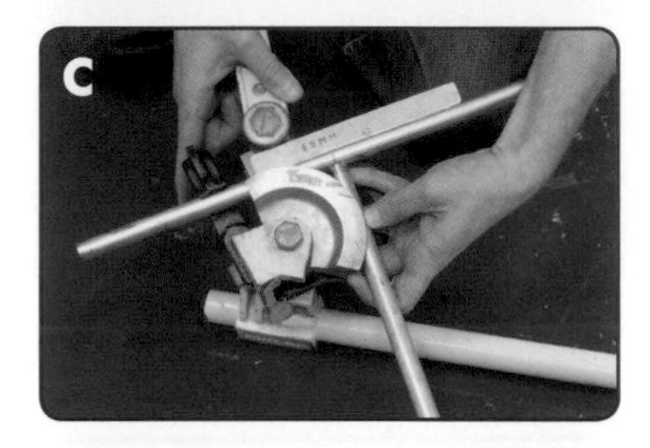

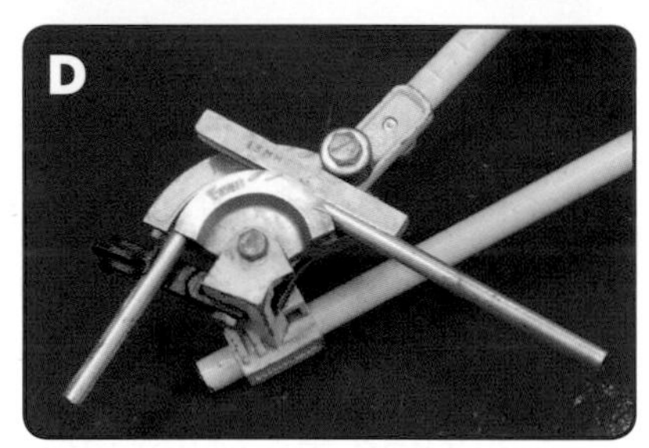

HAVE A GO...

Cut a piece of copper pipe at 200 mm, and bend it exactly in the centre at 90°.

Then measure from the centre of the bend to each end of the pipe and report your findings.

TRADE TIP

Wipe clean joints after soldering as any residues of flux could corrode the pipe.

Jointing copper pipe

Soldering

In order to carry out soldering, you will need the following items:

- Wire wool is used to clean the end of the pipe to make sure the solder sticks.
- Flux is applied to the end of cleaned pipe to prevent oxidation of the surface and help the solder to stick.

DID YOU KNOW

Pipes that carry hot or cold water must be soldered using lead-free solder, as leaded solder is poisonous if ingested.

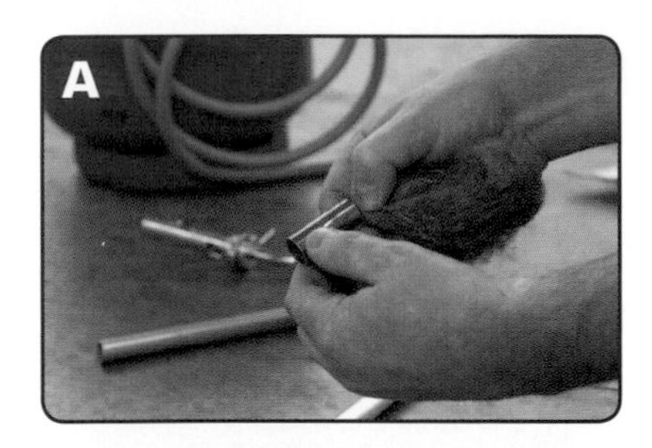

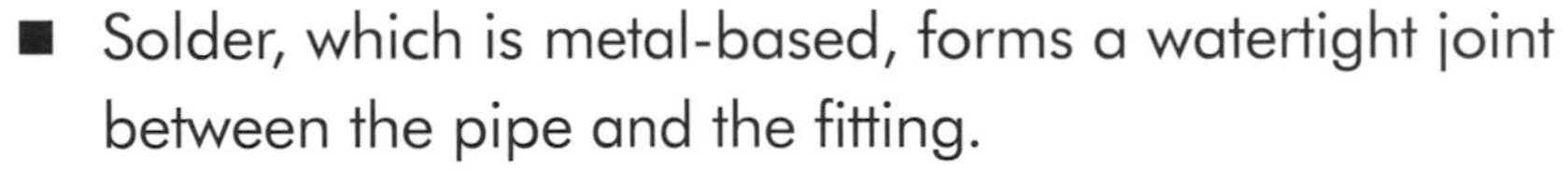

- Solder, which is metal-based, forms a watertight joint between the pipe and the fitting.
- A blow torch is used to heat the two surfaces and melt the solder.

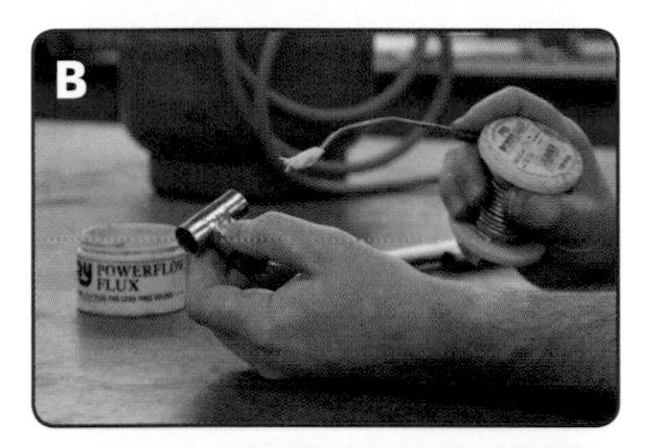

Capillary fittings

Jointing copper using capillary fittings takes lots of practice to perfect. It should be carried out as follows:

1. Clean the joints to be soldered with wire wool (**A**).
2. Apply flux evenly to the sections to be soldered (**B**). (Flux prevents the copper from oxidising.)

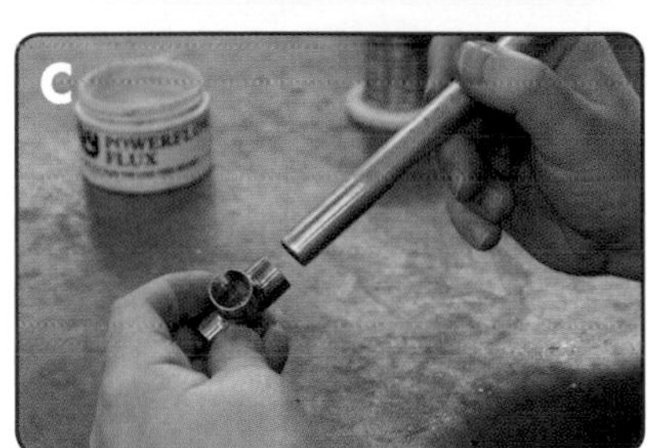

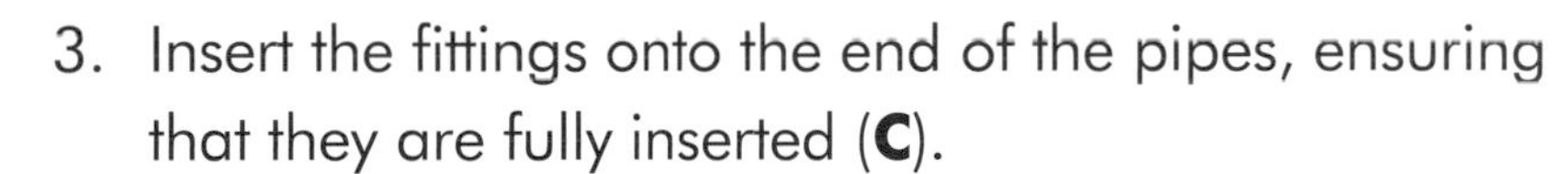

3. Insert the fittings onto the end of the pipes, ensuring that they are fully inserted (**C**).

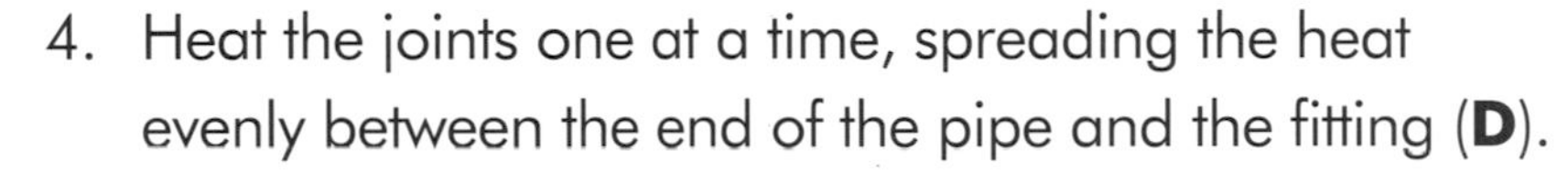

4. Heat the joints one at a time, spreading the heat evenly between the end of the pipe and the fitting (**D**).

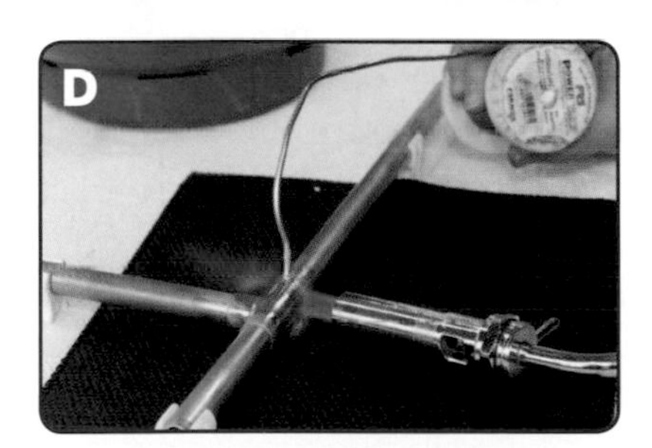

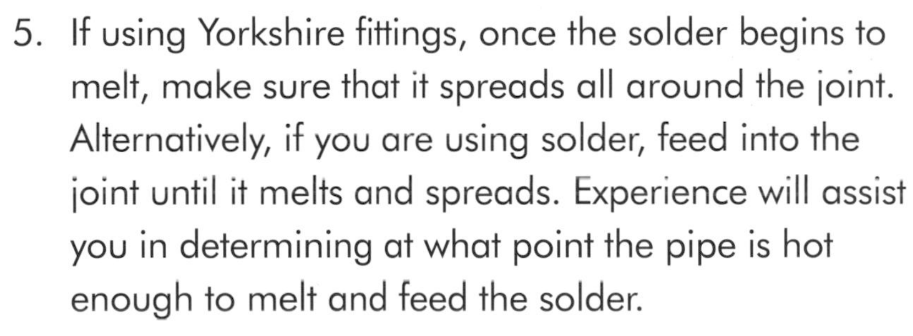

5. If using Yorkshire fittings, once the solder begins to melt, make sure that it spreads all around the joint. Alternatively, if you are using solder, feed into the joint until it melts and spreads. Experience will assist you in determining at what point the pipe is hot enough to melt and feed the solder.

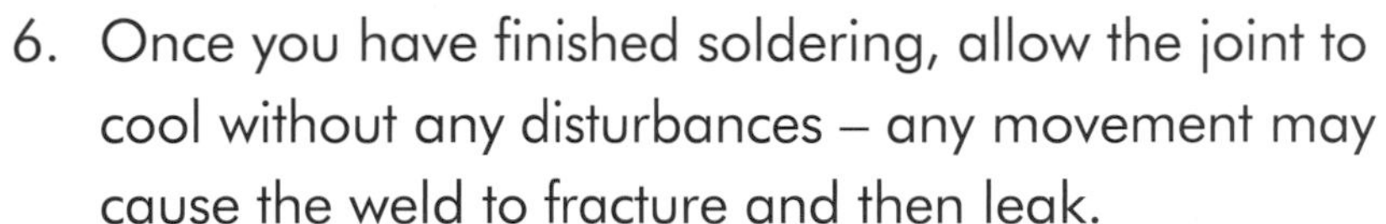

6. Once you have finished soldering, allow the joint to cool without any disturbances – any movement may cause the weld to fracture and then leak.

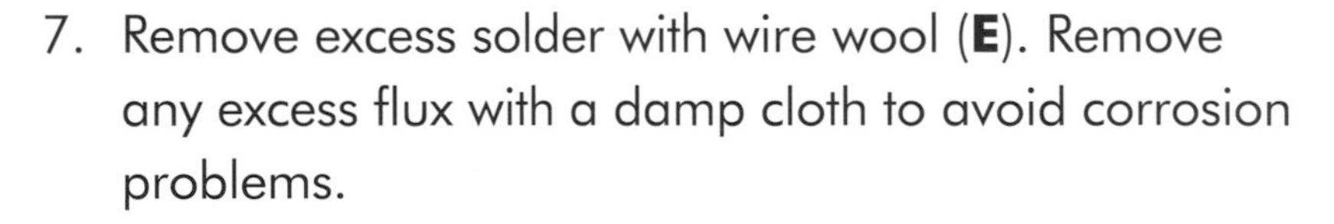

7. Remove excess solder with wire wool (**E**). Remove any excess flux with a damp cloth to avoid corrosion problems.

Solder two sections of copper pipe and a socket together

Interpreting drawings and specifications

Once you know how to cut, bend and solder pipe correctly, you will need to be able to interpret plumbing drawings in order to do the work correctly to specifications.

Fig. 9.24 is a simple drawing of some pipework that gives the type and size of pipe and fittings to use and their dimensions.

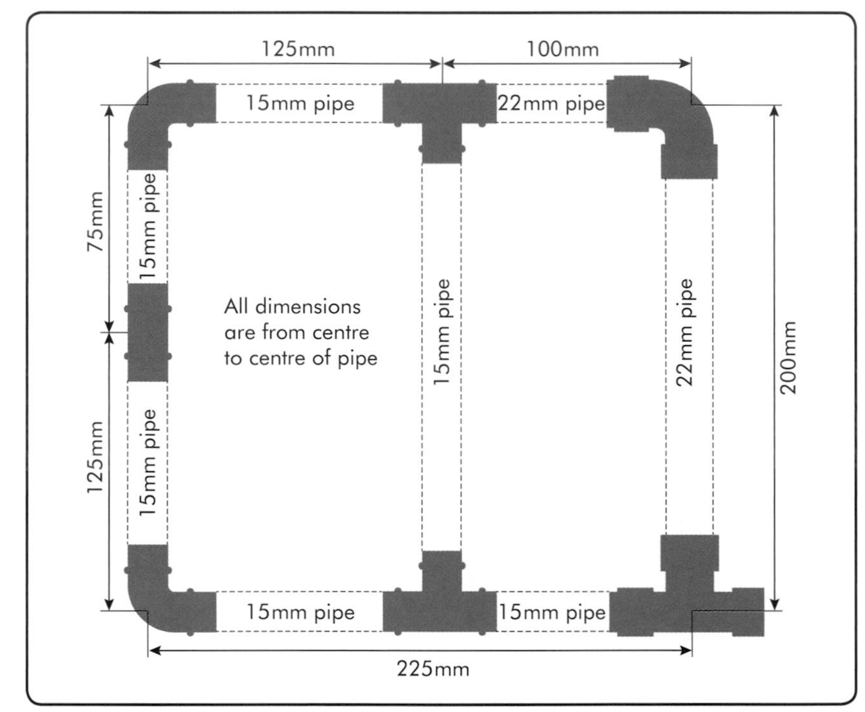

Fig. 9.24 Simple pipework diagram

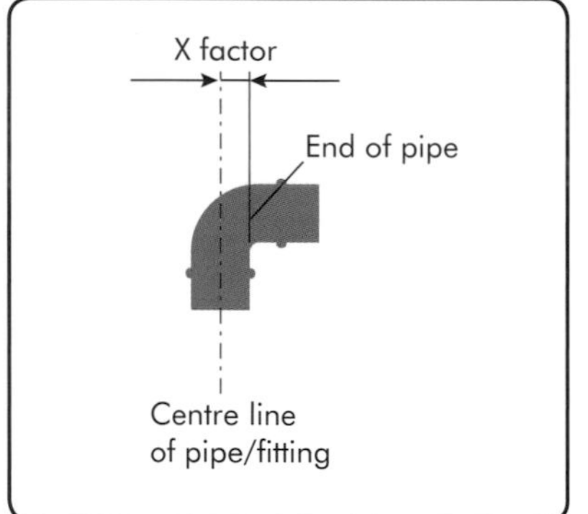

Fig. 9.25 The X factor is different for each fitting

Note that the measurements from centre to centre of the fittings do not correspond with the actual length of the pipe used. You need to allow for the X factor when measuring and cutting pipe.

Fabricating pipework

Before being able to fabricate pipework, you will need to master how to:

- cut, bend and solder copper pipework
- use compression and push-fit joints for both copper and plastic pipe
- choose the correct type of fixing for the pipe and the surface it is to be fixed to, and know how to fix it
- interpret drawings.

Once you have achieved this, you can start constructing simple pipework arrangements such as the one shown in Fig. 9.26.

How you fabricate the pipework will depend on its complexity. Typically, the process is as follows:

1. Determine if and where you would use pipe clips to keep the pipe in position.
2. If clips are required, mark the position of the clips on the wall or board using a spirit level and tape measure. Then, using the appropriate tools, fix the clips.
3. Look at the drawing or specification and determine the amount and size of pipe required.
4. From the drawing or specification, determine the necessary amount and size of fittings.
5. Determine the lengths of pipes required between each fitting, not forgetting to allow for the X factor and the pulled bends.
6. Cut and/or bend the required sections of pipe. (If the job is a big one, it is probably best to fabricate

REMEMBER

When measuring and cutting pipe for soldering, you will need to allow for the X factor of the fittings in order to avoid wastage.

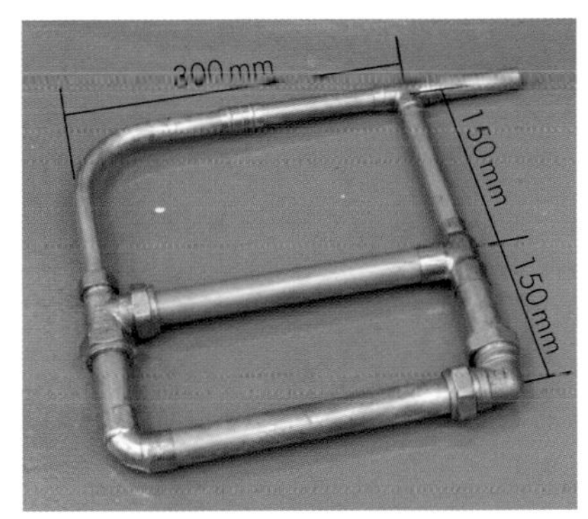

Fig. 9.26 Pipework

one section at a time, so cut and/or bend what is required for that section only.)

7. Once you have all your fittings and sections of cut or bent pipe, you should assemble them all together, but *without* applying flux to soldered fittings and pipe ends and jointing compound or PTFE tape to compression fittings. The reason this is done is to ensure that all cuts and bends have been made accurately – if you have made a mistake, then you will not mess up the end of the pipes or fittings.
8. If, once assembled, the fabrication is to specification, you can un-assemble and then re-assemble it, this time applying flux or jointing compound where required.
9. Any compression joints can be finger-tightened, making sure the pipe ends are fully inserted.
10. When soldering joints, be sure to protect the background from heat exposure using a heat-resistant mat. You must also be careful not to melt any plastic clips. If this cannot be achieved because the clip is very close to the joint, you may have to remove the clip and use something heat-resistant to hold the pipe in place. (A good tip is to use a brass clip temporarily in place of a plastic one.)
11. Once all joints have been soldered and have cooled down, wipe off all the flux with a damp cloth. This will prevent pipe corrosion.
12. Tighten all compression joints, making sure that all pipe ends are fully inserted. Use adjustable spanners and grips, if required, to hold the fitting in place so that it does not distort the pipe.

HAVE A GO...

Construct the pipework shown in Fig. 9.26 to the same dimensions and fix the pipework to the board.

Pressure testing pipework and fittings

All new pipework needs to be tested to make sure it does not leak. This is essential, especially when the pipework is then covered and may not be accessible in the future. If you do not test pipework, this could result in a leak causing major damage and an expensive repair bill.

All pipework has to be pressure tested to at least one and a half times the system working pressure. This is achieved using hydraulic testing equipment. At no stage should the pressure exceed 10 bar, as this could put the installation under too much stress and damage components, either in the installation or on the test equipment.

Copper or steel pipe should be pressure tested for 60 minutes. Plastic pipe needs to be tested to BS 6700 for plastic pipe, which is less stringent than metal pipe testing.

Fig. 9.27 Hydraulic pressure tester

TRADE TIP

Copper and plastic pipes have different pressure testing procedures, which can be found in the Water Supply (Water Fittings) Regulations (1999).

REMEMBER

When you have completed any plumbing activity, make sure that you leave your work area clean and tidy, return tools, equipment and excess materials, and dispose of any waste.

CHECK YOUR KNOWLEDGE

LEVEL 1

1. Name two items of PPE a plumber would use.
2. What would you carry out immediately after cutting pipe?
3. What type of solder is used on pipes that will carry cold water?
4. Which tool would you use to fix clips?
5. Name three different types of fittings.
6. What fixing device is commonly used for masonry walls?
7. Describe what the X factor is.
8. What pressure is pipework normally tested to?

Glossary

Accident book
This is required by law. Even minor accidents need to be recorded by the employer.

Alloy
A combination of two or more metals, which is designed to have greater strength or resistance to corrosion.

Alternating current (AC)
An electric current where the direction of current reverses at regular intervals.

Ambient temperature:
The temperature in a room, or the temperature that surrounds an object.

Anneal
A heating and cooling process that alters a material's properties.

Assembly point
An agreed place outside the building to head for in the event of an emergency.

Atom
The smallest unit of an element, made up of protons, neutrons and electrons.

Backflow
Waste or contaminated water entering the mains or freshwater supply, prevented by means of an air gap in the pipe system.

Blueprint
A design plan or technical drawing.

BSP
British standard pipe

Circuit breaker
A safety device that interrupts an electric current.

Commission
A process in which the installer checks the systems components and overall performance before it is handed over to the building owner and signed off.

Compound
A substance made up of two or more elements that are chemically joined together, for example, water is hydrogen and oxygen.

Conductor
Any material through which an electric current will flow.

Confined space
A working area that is substantially enclosed, in which accident or injury could occur.

Contact dermatitis
Inflammation of the skin following contact with a particular substance. The skin becomes red, dry, itchy and sore.

Contract specification
To formally agree upon the requirements of a job to be done.

COSHH
The Control of Substances Hazardous to Health Regulations are concerned with controlling exposure to hazardous materials.

Coulomb
The unit of electrical charge. 1 C is equivalent to a flow of 1 A in 1 S.

CPR
Cardiopulmonary resuscitation – a combination of rescue breaths and chest compressions to keep blood and oxygen circulating through the body.

Direct current (DC)
An electric current that flows in one direction only.

Eco-friendly
Having a minimal negative environmental impact.

Electron
A tiny particle that is part of an atom.

Ferrous
Metals that contain iron.

Fibre optic cable
A bundle of threads made of pure flexible glass, used instead of metal wiring.

Fluid substance
A gas or liquid, which can flow.

Fluorinated gas
A powerful greenhouse gas that contributes to global warming.

Flux
A substance that is used to prevent oxidation during soldering.

Force
The push or pull that acts between two objects.

Foundry
A place where metal is melted and poured into moulds.

Friction
Resistance to movement that is caused when one surface rubs against another.

Geothermal
Relating to the internal heat energy of the earth.

Gravity
The force of attraction between objects, which is very small, so is usually only felt for very large objects such as the earth.

Greywater
Waste water from washing machines, sinks and baths or showers.

HASAWA
The Health and Safety at Work etc. Act outlines your and your employer's health and safety responsibilities.

Hazardous waste
Waste that could cause great harm to people or the environment and needs to be disposed of with extreme care.

Hazard
Anything that could cause harm.

Heat sink
A heat exchanger that transfers heat from one source into a fluid, such as in refrigeration, air conditioning or the radiator in a car.

HSE
The Health and Safety Executive, an independent organisation that ensures health and safety laws are followed.

HVAC
The abbreviation for heating, ventilation and air conditioning.

Ion
An atom or molecule with a positive or negative charge.

Ionise
When atoms or molecules become charged by losing or gaining electrons.

Improvement notice
This gives the employer a time limit to make changes to improve health and safety.

Joule
The equivalent of passing 1 A of electric current through 1 Ω of resistance for 1 s.

Junction box
Contains terminals for joining electrical cables.

LCS
Low carbon steel tube. It is often referred to as black iron pipe or mild steel pipe or, if zinc coated, as galvanised iron pipe.

Low-risk waste
Waste that is not very dangerous but still needs to be disposed of carefully.

Malleable
Describes something that is flexible and can be bent without fracturing (breaking).

Molecular bond
A force that joins together the atoms within a molecule.

Molecule
Two or more atoms held together by strong chemical forces (bonds).

Moment
The turning effect of a force.

Non-ferrous
Metals that do not contain any iron.

Non-potable
Water that is unsuitable for drinking.

Olive
The more common name for the compression ring made from brass or copper that is used in a compression fitting.

Oxidise
A chemical process that adds oxygen to a substance.

Ozone layer
A thin layer of gas high in the earth's atmosphere.

Parallel circuit
Circuit in which components share the energy source but not the current.

pH
A measure of the acidity or alkalinity of a substance.

Photon
The basic unit of light or electromagnetic radiation.

PPE
Personal protective equipment can include gloves, goggles and hard hats.

Pressure
Force per unit area.

Principal contractor
This is either an individual working for 30 days or more on a construction job or 50 people working for 10 days or more.

Private contractor
A person or company that is temporarily employed by another business or an individual.

Prohibition notice
This stops all work until the improvements to health and safety have been made.

PVC
Polyvinyl chloride.

Recovery position
This involves rolling someone onto their side, with their head tilted back slightly to keep their airway open.

Renewable source
An energy source that is constantly replaced and will never run out, such as water, wind and solar energy.

Resistance
The way in which a material prevents the flow of electric current.

Respiratory protective equipment
Masks and breathing apparatus designed to prevent inhalation of harmful substances.

Retrofit
Adding new technology or features to something that already exists.

Risk
The chance, high or low, that somebody will be harmed by a hazard.

Risk assessment
An investigation that highlights the risks involved in a job and how to deal with those risks. The findings are recorded.

Self-employed
Someone who is in business for themselves and provides a service to a number of different customers or clients.

Semi-conductor
A substance that allows the passage of electricity in some conditions but not in others.

Series circuit
Circuit in which the components share the current.

Smart technology
Sensors that can detect whether a room or part of the building is occupied and then shut down systems, such as lighting, if the area is unoccupied.

Solder
Usually comes on a reel and is made from alloys of metal, such as tin or copper. It has a low melting point of about 450°C.

Stocks
The die-holding section of the tool.

Sub-contractor
An individual or group of workers that are directly employed by the contractor.

Superheat
When the temperature rises above boiling point. Superheated substances are in a vapour state.

Sustainability
In terms of building services engineering, this is about reducing a building's environmental impact over its lifetime.

Swarf
Small fragments of a metal produced when it is cut or filed.

Thermal insulator
Any material that is a poor conductor of heat.

Thermistor
A device in which electrical resistance changes with temperature.

Ultraviolet radiation
Electromagnetic radiation, which we cannot see but that has a heating effect (e.g. as sunburn).

VAT
Value Added Tax is charged on most goods and services. It is charged by businesses or individuals that have raised invoices in excess of £73,000 per year.

VDE
Verband der Elektrotechnik, the Association for Electrical, Electronic and Information Technologies.

Velocity
The speed of an object in a certain direction.

Z measurement
The distance between the centre of the fitting and the end of the tube when inserted.

Index